T0181798

Wireless Networks

Series Editor

Xuemin Sherman Shen

More information about this series at http://www.springer.com/series/14180

Peng Zhang • Chuang Lin

Security in Network Coding

 Springer

Peng Zhang
Department of Computer Science
   and Technology
Xi'an Jiaotong University
Xi'an, China

Chuang Lin
Department of Computer Science
Tsinghua University
Beijing, China

ISSN 2366-1186   ISSN 2366-1445   (electronic)
Wireless Networks
ISBN 978-3-319-80959-5   ISBN 978-3-319-31083-1   (eBook)
DOI 10.1007/978-3-319-31083-1

Printed on acid-free paper

This Springer imprint is published by Springer Nature
The registered company is Springer International Publishing AG Switzerland

Preface

Network coding is a new transmission paradigm proposed by Ahlswede et al. around the year 2000. It has been recognized as another breakthrough in information theory after Shannon. The key difference of network coding from traditional transmission paradigms is that in-network nodes can perform coding on packets, instead of simple store-and-forward. Benefits of network coding include higher data throughput, lower energy consumption, reduced bandwidth cost, etc. Due to these benefits, network coding has been applied to various kinds of networks, e.g., wireless mesh networks, content distribution networks, distributed storage networks, etc.

However, network coding also introduces new security and privacy challenges at the same time. The most serious problem is that it makes data transmission more vulnerable to *pollution attacks*. A single illegal packet can end up polluting a bunch of good ones through intermediate coding, and causing severe bandwidth waste. Besides data integrity, data confidentiality and user privacy also become quite different in the new context of network coding. On the one hand, we can leverage the intrinsic secrecy property of network coding to provide a lightweight confidentiality. On the other hand, we should redesign privacy-preserving mechanisms (e.g., anonymous routing) to make them compatible with network coding.

This book will first give a brief review of network coding in Chap. 1, including its benefits, applications, and security problems. Then, Chap. 2 will give a detailed review of security issues in network coding, highlighting how the security issues differ from those in traditional settings. In Chaps. 3–5, we will introduce three research works to address these security issues. The first work (Chap. 3) proposes a new method to defeat pollution attacks in network coding. Based on this method, a set of security mechanisms including a private key-based signature, a symmetric key-based MAC, and a hybrid key-based approach are presented and evaluated. The second work (Chap. 4) focuses on data confidentiality in network coding and presents a new encryption scheme named P-Coding. P-Coding recognizes the intrinsic secrecy property of random linear network coding and leverages it to offer a rather lightweight encryption which is very appealing in mobile ad hoc networks (MANETs). The third work (Chap. 5) identifies the problem that existing anonymous routing protocols (e.g., Tor) conflict with wireless network coding. This work introduces cooperation among wireless nodes in order to resolve such conflict, so that wireless user privacy and network coding benefits can be simultaneously maintained. Finally, in Chap. 6, we will conclude this book and discuss some future directions for the research "security in network coding."

Xi'an, China
Beijing, China

Peng Zhang
Chuang Lin

Acknowledgments

The authors would like to thank Xuemin (Sherman) Shen, Yixin Jiang, Patrick P. C. Lee, John C. S. Lui, Hongyi Yao, Chao Zhang, and Yanfei Fan for their contribution to the technical parts of this book.

Contents

Chapter 1
Introduction

Network coding [2] has been recognized as a significant improvement over the Shannon information transmission model. Many researchers in both information theory and computer communication areas have devoted a lot of effort to study network coding. In the following, we will give a brief review of network coding, followed by its benefits and applications. Then, we point out the security vulnerability of network coding, which is the focus of this book.

1.1 A Brief Review of Network Coding

The current Internet is designed and implemented based on the historical end-to-end principle proposed in 1984 [29]. This principle claims that intermediate nodes (e.g., switches and routers) should perform simple storing and forwarding on packets, without complicated operations; all the intelligence should be at end hosts. Network coding [2] can be thought as a challenge to this principle. The basic idea of network coding is to enable intermediate nodes to perform coding on packets they receive, rather than just simply forwarding them. In this sense, data transmitted in the network are no longer immutable packets but encoded packets that are continuously processed by nodes.

To further illustrate the basic idea of network coding, consider Fig. 1.1 as an example. In traditional store-and-forward transmission paradigm, when a router receives a packet from link e_i^{in}, it looks up in the routing table for the output link, say e_j^{out}. Then, the router outputs the packet with the payload intact to the link e_j^{out}. For network coding, when a router receives multiple packets from links, say $e_1^{in}, e_2^{in}, e_3^{in}$, it applies a coding function $f_{e_j^{out}}$ on the received packets and outputs the resultant packet to an output link e_j^{out}. For different output links e_j^{out}, the function $f_{e_j^{out}}$ would be different.

© Springer International Publishing Switzerland 2016
P. Zhang, C. Lin, *Security in Network Coding*, Wireless Networks,
DOI 10.1007/978-3-319-31083-1_1

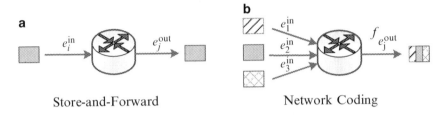

Fig. 1.1 Traditional store-and-forward paradigm vs. network coding

Depending on different applications, the coding function can be linear or nonlinear, deterministic, or probabilistic. For example, in random linear network coding [14], the coding function is defined as a random linear function, i.e., nodes linearly combine input packets with randomly chosen coefficients.

1.2   Benefits of Network Coding

In network coding, packets in the network become the carrier of "information flows," rather than "commodity flows." From information-theoretical point of view, information's flow style of transmission can greatly tap the full potential of communication networks and achieve a higher data throughput [2, 16, 20, 22]. Besides throughput, previous researches on network coding have shown that network coding can reduce transmission energy consumption [10, 21, 31], simplify transmission control [11, 30], and minimize link bandwidth cost [7]. In the following, we will give a brief review of these benefits:

- **Higher throughput.** In multicast scenarios, suppose the maximum flow or minimum cut from the source s to each sin t is $MinCut(s, t)$, then the multicast capacity is $C = min_{t \in T} MinCut(s, t)$. That is, s can only transmit C units of information to each sink t. This capacity cannot be achieved with store-and-forward transmission but can be obtained when network coding is used. In addition, researchers have proven that the codes for achieving multicast capacity can be calculated in polynomial time [2, 16, 20, 22].
- **Lower energy consumption.** In wireless networks, each node can work in promiscuous mode to overhear transmissions of neighboring nodes. Leveraging this property, wireless nodes can opportunistically broadcast XORed packets that they want to send [19]. Nodes that received the XORed packets can decode packets that are sent to them, by decoding the packets that they have overheard. This can reduce the transmission times in wireless networks and thus reduce the energy consumption of wireless nodes [10, 21, 31].
- **Easier transmission control.** In P2P content distribution networks, peers need to exchange control messages with each other to know what blocks that each peer holds. With this information, each node can retrieve different blocks from

the corresponding peers, thereby avoiding redundant block transmissions. The control messages can be saved with random linear network coding: each peer generates random linear combinations of blocks that they hold and randomly selects peers to exchange blocks [11, 30]. Since any random linearly coded blocks are different with high probability, each peer can obtain the content once he/she has retrieved enough coded blocks.

- **Lower bandwidth cost.** Erasure codes are commonly used in distributed storage systems, so that when a storage node fails, the content can be recovered/repaired from other nodes. Network coding can act as a special kind of erasure code and be used for distributed storage. It has been proven that network coding can minimize the bandwidth cost when recovering the data of failed nodes [7].

Apart from the above benefits, network coding has also been shown to improve system reliability [25, 26] and reduce transmission latency [6].

1.3 Applications of Network Coding

Since its birth in 2000, network coding has found many applications in wired and wireless networks.

First, in wired networks, Gkantsidis et al. [11] leverage network coding to simplify the scheduling of data block transmission in content distribution networks (CDNs). Meanwhile, the authors show that the file downloading time can be reduced by 20–30 %, compared to traditional source coding. Wang et al. [30] propose a new mode of live streaming termed "random push with random network coding" or R^2. Liu et al. [24] analyze the performance improvement of UUSee video on demand (VoD) service with network coding.

Second, in wireless networks, Katti et al. [19] propose wireless network coding, e.g., COPE. In COPE, nodes work in promiscuous mode and monitor packets sent by their neighbors. When a node finds a coding opportunity, it calculates the XOR of the sending packets and broadcasts the XORed packet. Nodes that receive this XORed packet use their overheard packets to decode it, in order to obtain the packets that are destined for them. Later, Katti et al. [18] propose symbol-level network coding, which can be used for live streaming in vehicular networks. Chachulski et al. [5] propose MORE, a new opportunistic routing protocol based on random linear network coding. MORE tries to mitigate the complexity of node scheduling in ExOR [4], by letting nodes forward random linear combinations of overheard packets. Due to randomness, packets sent by each node will be different with high probability, and thus there is no redundant transmission. Also for wireless networks, Zhang et al. [34] introduce physical-layer network coding. Compared with COPE, this coding method can further reduce the transmission times be leveraging the interference of packets.

Other applications of network coding include network tomography [8, 13, 23, 32], distributed data storage [1, 7, 15], secure key distribution [27, 33], etc. Interested readers can refer to [9] for more details.

As we know, computer networks are prone to various kinds of attacks, including DDoS [28] and prefix hijacking [3]. In addition, information security and privacy are also vulnerable to threats like traffic eavesdropping, Byzantine modifications, packet replays, etc. For networks that enable network coding, they may still be faced with the above security threats, some of which may cause even worse consequences. In addition, network coding also brings some new threats that are absent in traditional networks. Even worse, existing methods cannot effectively prevent such threats.

1.4 Insecurity of Network Coding

Though network coding is very appealing, it also introduces new security problems. If not addressed well, these problems can potentially limit the adoption of network coding in real networks. In the following, we will give a brief review of the security problems in network coding:

- Network coding is vulnerable to pollution attacks [12]. A packet modified by an adversary will encode with other good packets and thus results in many more packets being "polluted." The polluted packets will prevent receivers from decoding the correct messages. Thus, pollution attacks can cause a great waste of network bandwidth and serious degradation of system availability. In addition, work coding is also vulnerable to a new kind of attack named *entropy attack* [17]. In entropy attack, an adversary does not modify any packet but injects a lot of legal but "non-innovative" packets into the network. By non-innovative packets, these are linearly dependent on each other and thus cannot help the receivers decode the source packets. Entropy attack will lower the entropy of coded packets transmitted in the network and reduce the success probability of sink decoding.
- Network coding makes the problem of data confidentiality different from that in traditional networks. On the one hand, packets transmitted in the network are no longer plaintexts but encoded ones. Thus, network coding offers an intrinsic confidentiality. On the other hand, the adversary can still obtain the plaintexts by collecting enough number of packets, and decoding them via Gaussian eliminations. Therefore, network coding itself cannot provide enough content confidentiality and is still vulnerable to eavesdropping attacks.
- Wireless network coding leaves user privacy at risks. In traditional networks, the attacker can monitor the traffic of a wireless node and trace the originator of packets by correlating the content, length and sending/receiving time of packets. Existing methods like Onion Routing use layered encryption/decryption to eliminate the correlation of packets and thus can defeat the above attacks. However, in wireless network coding, the layered encryption/decryption is conflicting with network coding operations, making all anonymity schemes based on Onion Routing infeasible.

Noted from the above security problems, we set out to introduce various countermeasures in this book. We will first present schemes proposed by others in the first (Chap. 2) and then introduce three of our solutions in Chaps. 3, 4, and 5.

1.5 Book Organization

The rest of this book is organized as follows:

- Chapter 2 will give a more detailed introduction to the security problems in network coding, including pollution attacks, eavesdropping attacks, entropy attacks, and privacy problem. The corresponding countermeasures to these attack will also be discussed.
- Chapter 3 will introduce a method, termed "padding for orthogonality," to defeat pollution attacks in network coding. Based on this method, two concrete constructions, e.g., homomorphic subspace signature (HSS) and homomorphic subspace MAC (HSM) will be presented. After that, another construction named MacSig will be introduced, which combines the advantages of HSS and HSM. Finally, this chapter will analyze the bandwidth and computation overhead of MacSig and compare it with other signature-based schemes.
- Chapter 4 will center around data confidentiality in network coding and introduce P-Coding, a secure network coding scheme for MANET. This chapter will first give an analysis on the intrinsic security property of random linear network coding. Then, P-Coding is introduced as a lightweight security scheme for network coding. This chapter will validate the security of P-Coding through both analysis and experiments and compare the performance of P-Coding with other security schemes.
- Chapter 5 will present ANOC, a new anonymous network-coding-based routing protocol. This chapter begins by outlining the problem that traditional anonymity schemes conflict with wireless network coding and then shows how to solve this problem by enabling cooperations among wireless nodes. Then, this chapter will introduce the detailed design and implementation of ANOC and present the experimental results on ANOC.
- Chapter 6 will first give a concluding remarks for this book and then outline some future research directions for security in network coding.

References

1. Acedanski, S., Deb, S., Médard, M., Koetter, R.: How good is random linear coding based distributed networked storage. In: Proceedings of International Workshop on Network Coding, Theory and Applications (NetCod) (2005)
2. Ahlswede, R., Cai, N., Li, S.Y., Yeung, R.W.: Network information flow. IEEE Trans. Inf. Theory **46**(4), 1204–1216 (2000)

3. Ballani, H., Francis, P., Zhang, X.: A study of prefix hijacking and interception in the internet. ACM SIGCOMM Comput. Commun. Rev. **37**(4), 265–276 (2007)
4. Biswas, S., Morris, R.: Opportunistic routing in multi-hop wireless networks. ACM SIG-COMM Comput. Commun. Rev. **34**(1), 69–74 (2004)
5. Chachulski, S., Jennings, M., Katti, S., Katabi, D.: Trading structure for randomness in wireless opportunistic routing. ACM SIGCOMM Comput. Commun. Rev. **37**(4), 169–180 (2007)
6. Chou, P.A., Wu, Y.: Network coding for the internet and wireless networks. IEEE Signal Process. Mag. **24**(5), 77–85 (2007)
7. Dimakis, A.G., Godfrey, P.B., Wu, Y., Wainwright, M.J., Ramchandran, K.: Network coding for distributed storage systems. IEEE Trans. Inf. Theory **56**(9), 4539–4551 (2010)
8. Fragouli, C., Markopoulou, A.: A network coding approach to network monitoring. In: Proceedings of the 43rd Allerton Conference on Communication, Control, and Computing (2005)
9. Fragouli, C., Soljanin, E.: Network Coding Applications. Now Pub, Hanover (2008)
10. Fragouli, C., Widmer, J., Boudec, J.: A network coding approach to energy efficient broadcasting: from theory to practice. In: Proceedings of IEEE INFOCOM (2006)
11. Gkantsidis, C., Rodriguez, P.R.: Network coding for large scale content distribution. In: Proceedings of IEEE INFOCOM, pp. 2235–2245 (2005)
12. Gkantsidis, C., Rodriguez, P.: Cooperative security for network coding file distribution. In: Proceedings of IEEE INFOCOM, vol. 6, pp. 1–13 (2006)
13. Ho, T., Leong, B., Chang, Y.H., Wen, Y., Koetter, R.: Network monitoring in multicast networks using network coding. In: Proceedings of International Symposium on Information Theory (ISIT), pp. 1977–1981. IEEE, Adelaide (2005)
14. Ho, T., Médard, M., Koetter, R., Karger, D.R., Effros, M., Shi, J., Leong, B.: A random linear network coding approach to multicast. IEEE Trans. Inf. Theory **52**(10), 4413–4430 (2006)
15. Hu, Y., Chen, H.C., Lee, P.P., Tang, Y.: Nccloud: applying network coding for the storage repair in a cloud-of-clouds. In: Proceedings of USENIX FAST (2012)
16. Jaggi, S., Sanders, P., Chou, P.A., Effros, M., Egner, S., Jain, K., Tolhuizen, L.M.: Polynomial time algorithms for multicast network code construction. IEEE Trans. Inf. Theory **51**(6), 1973–1982 (2005)
17. Jiang, Y., Fan, Y., Shen, X., Lin, C.: A self-adaptive probabilistic packet filtering scheme against entropy attacks in network coding. Comput. Netw. **53**(18), 3089–3101 (2009)
18. Katti, S., Katabi, D., Balakrishnan, H., Medard, M.: Symbol-level network coding for wireless mesh networks. ACM SIGCOMM Comput. Commun. Rev. **38**(4), 401–412 (2008)
19. Katti, S., Rahul, H., Hu, W., Katabi, D., Médard, M., Crowcroft, J.: Xors in the air: practical wireless network coding. ACM SIGCOMM Comput. Commun. Rev. **36**(4), 243–254 (2006)
20. Koetter, R., Médard, M.: An algebraic approach to network coding. IEEE/ACM Trans. Netw. **11**(5), 782–795 (2003)
21. Li, L., Ramjee, R., Buddhikot, M., Miller, S.: Network coding-based broadcast in mobile ad-hoc networks. In: Proceedings of IEEE INFOCOM (2007)
22. Li, S.Y., Yeung, R.W., Cai, N.: Linear network coding. IEEE Trans. Inf. Theory **49**(2), 371–381 (2003)
23. Lin, Y., Liang, B., Li, B.: Passive loss inference in wireless sensor networks based on network coding. In: Proceedings of IEEE INFOCOM, pp. 1809–1817 (2009)
24. Liu, Z., Wu, C., Li, B., Zhao, S.: Uusee: Large-scale operational on-demand streaming with random network coding. In: Proceedings of IEEE INFOCOM, pp. 1–9 (2010)
25. Lun, D.S., Médard, M., Koetter, R., Effros, M.: Further results on coding for reliable communication over packet networks. In: Proceedings of International Symposium on Information Theory (ISIT), pp. 1848–1852 (2005)
26. Lun, D.S., Médard, M., Koetter, R., Effros, M.: On coding for reliable communication over packet networks. Phys. Commun. **1**(1), 3–20 (2008)
27. Oliveira, P.F., Barros, J.: A network coding approach to secret key distribution. IEEE Trans. Inf. Forensics Secur. **3**(3), 414–423 (2008)

28. Park, K., Lee, H.: On the effectiveness of route-based packet filtering for distributed dos attack prevention in power-law internets. ACM SIGCOMM Comput. Commun. Rev. **31**(4), 15–26 (2001)
29. Saltzer, J.H., Reed, D.P., Clark, D.D.: End-to-end arguments in system design. ACM Trans. Comput. Syst. **2**(4), 277–288 (1984)
30. Wang, M., Li, B.: R2: Random push with random network coding in live peer-to-peer streaming. IEEE J. Sel. Areas Commun. **25**(9), 1655–1666 (2007)
31. Wu, Y., Chou, P.A., Kung, S.Y.: Minimum-energy multicast in mobile ad hoc networks using network coding. IEEE Trans. Commun. **53**(11), 1906–1918 (2005)
32. Yao, H., Jaggi, S., Chen, M.: Network coding tomography for network failures. In: Proceedings of IEEE INFOCOM, pp. 1–5 (2010)
33. Zeng, R., Jiang, Y., Lin, C., Fan, Y., Shen, X.S.: A scalable and robust key pre-distribution scheme with network coding for sensor data storage. Comput. Netw. **55**(10), 2534–2544 (2011)
34. Zhang, S., Liew, S.C., Lam, P.P.: Hot topic: physical-layer network coding. In: Proceedings of the 12th Annual International Conference on Mobile Computing and Networking, pp. 358–365 (2006)

Chapter 2
Security Threats in Network Coding

While network coding can help improve network performance, it also introduces new security and privacy issues. For example, network coding can make data transmission more vulnerable to "pollution attacks": A single illegal packet can end up polluting a bunch of good ones through the process of intermediate coding, causing receivers unable to decode properly. Besides data integrity, data confidentiality and user privacy also become quite different in the new environment of network coding. This makes existing schemes like digital signatures, encryption algorithms, and anonymity schemes either infeasible or inefficient. This chapter will introduce several attacks on network coding and the corresponding countermeasures.

2.1 Pollution Attack

Among all the threats of network coding considered so far, pollution attacks are perhaps the most concerned ones. Figure 2.1 gives an example of pollution attack in network coding. We can see that a single polluted packet can end up polluting many more good ones. To thwart this kind of attacks, many schemes have been proposed. We classify them into two categories: *information-theoretic schemes* and *cryptography-based schemes*.

2.1.1 Information-Theoretic Schemes

Information-theoretic schemes mostly leverage error correction codes to add redundancy to the messages at source nodes. In this way, destination nodes can recover the original messages from the received packets (which may contain polluted packets) through decoding. Methods proposed in [9, 49, 53] can correct the transmission

© Springer International Publishing Switzerland 2016
P. Zhang, C. Lin, *Security in Network Coding*, Wireless Networks,
DOI 10.1007/978-3-319-31083-1_2

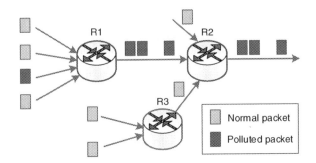

Fig. 2.1 Example of pollution attack in network coding

errors in network coding. They are not designed to detect pollution attack; they can only mitigate the attack to some extent. Ho et al. [21] propose an extension to random linear network coding to detect Byzantine modification (i.e., pollution attack), by adding hashes in the source messages. Jaggi et al. [23] propose three different polynomial-time source coding methods. Suppose the communication capacity is C and the bandwidth of the attack is z_o. Then, a secure throughput of $C - z_O$ or $C - 2z_O$ can be achieved using these codes.

Kotter et al. [28] model the error correction of network coding as the problem of transmitting a *vector space* through an operator channel. The authors observe that the only invariant of error-free random linear network coding is the linear space spanned by packets. By checking whether the vector space generated by the received packets is the same as that generated by the source packets, the receiver can detect whether there is a transmission error or malicious modification. Formally, suppose the source transmits a linear subspace V, each of whose basis corresponds to a source packet. Since V is closed under linear combinations, it will remain the same after network coding. Similarly, the destination also constructs a linear subspace U from all received packets. If Rank($V \cap U$) is sufficiently large, the transmission can be said to be error-free. Based on this idea, the authors introduced a new metric termed *rank distance* and proposed an optimal coding scheme similar to Reed-Solomon codes [47]. Different from [23], this coding scheme can operate on any finite field and has no restriction on packet size.

2.1.2 *Cryptography-Based Schemes*

To actively prevent pollution from propagating among intermediate nodes, several *cryptography-based schemes* have been proposed. They can be further grouped into two classes: *public key cryptographic approaches* and *symmetric key cryptographic approaches*

Public-Key Cryptographic Approaches In the innovative work of Krohn et al. [30], *homomorphic hash function* is proposed to enable on-the-fly verification for erasure codes. As the verification process requires nodes to compute expensive homomorphic hash functions, the technique of *batched verification* is employed. Gkantsidis et al. [20] extend this scheme to network coding-based P2P networks and further reduce the computation overhead by enabling cooperative verification among peers. One common limitation of these two schemes is that the hash values of the whole file should be computed at the source and delivered to downstream nodes in advance.

To address the above problem, Yu et al. [50] propose a *homomorphic signature* scheme on basis of homomorphic hash function and RSA cryptosystem. In their scheme, each packet carries an RSA-encrypted homomorphic hash, which functions as its homomorphic signature. A recent work [52], however, shows that this signature scheme is only conditionally valid and may be vulnerable to trivial no-message attacks. Charles et al. [10] introduce a secure homomorphic signature based on Weil pairing over elliptic curves [6], which is even more expensive than homomorphic hash functions.

From quite a different perspective, Zhao et al. [54] propose a signature scheme in which relay nodes check the integrity of packets by verifying whether they belong to the subspace of source packets. Boneh et al. [7] introduce another signature scheme similar to [54], but use signatures of a smaller size. However, both [54] and [7] still require extra secure channels to pre-distribute signatures, just like [20, 30]. To sum up the above public key cryptographic approaches, they are not computationally efficient due to the expensive operations of homomorphic hashing or signatures.

Symmetric Key Cryptographic Approaches In the scheme proposed by Yu et al. [51], the source attaches to each packet multiple MACs, each of which authenticates part of the packet. These MACs are encrypted using different keys at the source, and relay nodes can cooperatively check different MACs using their respectively shared keys. Agrawal et al. [2] propose a similar MAC-based scheme, which differs from [51] in that each MAC authenticates the whole packet. However, this scheme is unfortunately vulnerable to tag pollution, which can also degrade the system performance.

Toward this problem, Li et al. [33] propose RIPPLE, a time-based authentication protocol based on TESLA [38]. The key idea of RIPPLE is time asymmetry: the source utilizes a stream of keys to generate multiple tags for each packet, but does not pre-distribute these keys. The keys are distributed only when nodes need to verify a packet. In this way, an adversary does not know the keys being used in prior and thus cannot forge valid tags. Similarly, Dong et al. [16] propose another TESLA-based method named DART. In DART, the source node continuously generates random MACs after it has sent packets. Other nodes only use the MACs that are generated "after" the time when they received the packets for verification. A common problem with RIPPLE and DART is that they require global synchronization among all nodes, which is not easy to be realized in distributed settings. In addition, since nodes need to wait for the keys or tags, they will incur an additional transmission latency.

Kehdi et al. [27] propose another symmetric key-based scheme, which utilizes *null keys* for verification. A null key is just a vector from the null space of the matrix formed by the source packets; legitimate packets are supposed to map null keys to zero. This scheme may incur a high bandwidth overhead since it injects into the network with multiple null keys for each generation. In sum, the above symmetric key cryptographic approaches are quite efficient in computation, but they must have carefully manage the keys and incur a relatively large bandwidth overhead.

They are many other schemes for defending against pollution attacks [25, 31, 32], and we will not cover them for limited space.

2.2 Eavesdropping Attack

There are several works focusing on secure network coding against eavesdropping attacks. We broadly group them into the following three classes according to the security level that they aim to achieve:

2.2.1 Shannon Security

The first category of works attempt to provide *Shannon security* by designing secure linear codes for network coding. Cai et al. [8] first identify the wiretap adversary who can monitor a limited number of links and present an approach to transform any linear network code to be secure using secret sharing. More specifically, suppose the max-flow capacity of the network is n, and the adversary can monitor at most k edges; then the sender should send $n - k$ message symbols with another k random symbols, so that no information will be leaked to the adversary. However, Cai's scheme is not very efficient as it requires a large filed size and takes a large number of steps.

This problem is treated by Feldman et al. [19], who generalize the method used in [8] and find a trade-off between the multicast capacity and the field size. In this way, the filed size can be significantly reduced by giving up a small amount of capacity. An interesting observation made in [19] is that making network coding secure is equivalent to finding codes with some distance properties. Rouayheb et al. [17] formalize the problem as a generalization of Ozarow–Wyner wiretap channel II model [37] and propose a secure scheme by implementing cosset coding at the source. They show that this coding scheme can be implemented without affecting the underlying network coding architecture. This model is further studied by Silva et al. in [44], where another source coding scheme, named maximum rank distance (MRD), is designed to replace the codes used in [17]. Both [17] and [44] stress that the designing of secure codes and the optimizing of network transport can be treated independently.

Different from the above approaches, Adeli et al. [1] propose a different secure network coding scheme with small overhead by utilizing hash functions. In their scheme, each generation includes one uniformly distributed symbol as a secret, and the randomness of this symbol is scattered into other data symbols using a hash chain. This scheme is claimed able to provide complete security for linear network coding, regardless of the number of independent coding vectors acquired by the adversary. However, a more extensive proof is needed to justify this argument. To sum up the above schemes, they all assume that the eavesdroppers are limited in listening capability and unnecessarily sacrifice certain capacity to achieve Shannon security.

2.2.2 Weak Security

Bhattad et al. [4] introduce the concept of *weak security*, by which the system is said to be secure if the adversary cannot recover any "meaningful" information. To illustrate this concept, consider Fig. 2.2, for example. In Fig. 2.2a, x_1 is the message to be transmitted, and w is a randomly generated key. In order to recover the source message x_1, the attacker should at least eavesdrop at two different links. Then, the transmission is Shannon secure given the adversary can only eavesdrop on one link. In Fig. 2.2b, x_1 and x_2 are the source messages to be transmitted. An adversary that eavesdrops on one link can still obtain some information (linear combinations of x_1 and x_2). In this sense, the transmission is not Shannon secure. However, the attacker cannot decode either x_1 or x_2 and thus cannot recover only "meaningful" information about them. Then, the transmission is said to be weakly secure. As seen in Fig. 2.2b, the source can transmit two messages simultaneously, meaning that weak security condition permits a higher transmission throughput compared

Fig. 2.2 Comparison of Shannon security and weak security [4]. (**a**) Shannon secure transmission. (**b**) Weakly-secure transmission

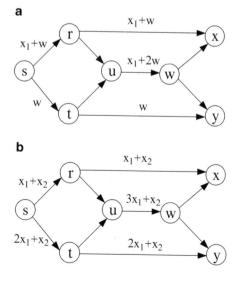

to Shannon security. The authors also show that the weak security and multicast capacity can be simultaneously achieved, by performing linear transformations at the source. They also show that random linear network coding is inherently weakly secure with a high probability if coding is performed over a large finite field.

2.2.3   Computational Security

Lima et al. [34] consider the threats posed by "nice but curious" intermediate nodes and develop an algebraic security criterion to access the intrinsic security provided by network coding. They derive the relationship between field size and the security level and observe that the security is dependent on network topology. The algebraic security criterion is essentially weak security. Based on the weak security model, Wang et al. [46] design a polynomial-time deterministic code to secure linear network coding. They show that by using this scheme, optimal throughput for multiple streams between a single-source destination pair can be achieved.

By leveraging the intrinsic security of network coding, some cryptographic approaches have been proposed to secure network coding-based applications. One scheme is SPOC [35, 45], proposed by Vilela et al., in which the source encrypts/locks the GEV of each message after random linear coding and attaches another set of GEVs to enable standard network coding. Receivers can recover the source messages by following a decode-decrypt-decode procedure. This scheme is essentially an end-to-end cryptographic approach and is lightweight in computation. This scheme requires the source performs the following four steps before transmission: (1) prefix each message with the corresponding unit vector acting as coding vector; (2) linearly combine the prefixed messages using random coefficients; (3) perform symmetric encryption on the coding vector of each message; and (4) attach another coding vector as prefix for each message. After these four steps, the messages are sent, and intermediate nodes perform standard network coding on this packet. Sinks can recover the source messages by following a decode-decrypt-decode procedure.

Another scheme proposed by Fan et al. [18] is based on homomorphic encryption function (HEF) [3]. This scheme has the coding coefficients encrypted using HEF. Due to the homomorphic property of HEF, linearly combined operations can be directly performed on the encrypted coding coefficients. As a result, no extra coding coefficients are needed as by SPOC. As another difference from SPOC, Fan's HEF-based scheme can achieve both content secrecy (i.e., confidentiality) and contextual secrecy (i.e., privacy) at the same time. However, both of these two schemes fail to fully exploit the mixing nature of network coding.

2.3 Entropy Attack

Entropy attack [24] is a new DoS attack that is specially targeted at network coding. In entropy attack, the adversary injects packets that are legal, meaning that these packets are linear combinations of source packets. However, these packets cannot increase the rank of the subspace generated by received packets and thus are not useful for receivers to decode. This attack gets its name since the entropy of the coded packets becomes low when the attack is launched. Entropy attack can cause a waste of bandwidth resource and reduce the probability of successful decoding at receivers.

Take Fig. 2.3 as an example. The source S tries to send some packets to the destination D. An adversarial node R_1 residing in the network, R_2 outside the network tries to launch entropy attacks. Suppose at each time slot t_i, the source S sends two coded packets $y_{t_i}(e_1)$ and $y_{t_i}(e_2)$ through two output links e_1 and e_2, respectively. R_1 continuously uses packets $y_{t_1}(e_1)$ and $y_{t_2}(e_1)$ to generate "non-innovative" packets $y_{t_{3+}}(e_3) = ay_{t_1}(e_1) + by_{t_2}(e_1)$ and sends them to destination D. Similarly, R_2 can also generate and inject "non-innovative" packets $y_{t_{2+}}(e_3') = a'y_{t_1}(e_1) + b'y_{t_1}(e_2)$ with intercepted packets. In this way, the destination D will not receive enough innovative packets and cannot decode the source packets.

In order to prevent the entropy attack, we can let nodes check the linear dependence of their received packets, and a new packet that is a linear combination of already received packet should be discarded. However, since the finite field size and packet size are relatively large, checking linear dependence will bring high computation load to nodes. To address this problem, Jiang et al. [24] propose a probabilistic algorithm that can detect and filter non-innovative packets fast. The authors show that when the finite filed size is 2^{256} and the packet length is 384, the computation overhead is only 4 % of the traditional linear dependence checking cost.

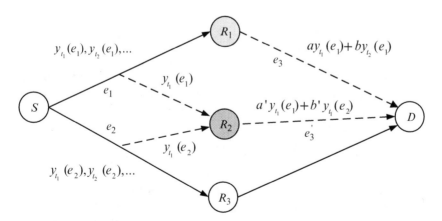

Fig. 2.3 Example of entropy attack in network coding [24]

2.4 Privacy Leakage

User privacy is a great concern in today's Internet. For wireline networks, there are two fundamental techniques, i.e., Mix-Net and Onion Routing. Here, Mix-Net refers to schemes that are based on Chaum's mix [11], e.g., Mixminion [12], Mixmaster [36] and MorphMix [41, 42]. The common feature of them is that they all employ techniques such as *shaping*, *reordering*, and *layered encryption* to eliminate the packet correlations at participating nodes. By layered encryption, the source should successively encrypt each packet with public keys of the nodes along the route. Then, each node peels one layer of encryption with its private key so that the packet finally arrives at the receiver as plaintext.

Onion Routing [39] refers to a family of anonymity protocols, which are also based on the technique of layered encryption, but is more computationally efficient than Mix-Net. In traditional Onion Routing, the source creates a layered structure named *onion*, by successively encrypting session keys for nodes along the route using their corresponding pubic keys. Then, each node along the route decrypts the onion with its private key to obtain the session key for it. After that, data packets are moved along the route just as in Mix-Nets, except that here packets are symmetrically encrypted by the source with the session keys previously distributed. The technique of Onion Routing is further developed in Tor [15], which ensures forward secrecy for Onion Routing using incremental path building, and in [26], which eliminates the need of PKI in Onion Routing using multi-path routing. Based on Tor, DiBenedetto et al. propose ANDaNA [14], which can provide anonymity for content centric networks [22].

As noted above, both Mix-Net and Onion Routing require relay nodes to perform encryptions/decryptions on packets, and these operations are in conflict with the packet-mixing operations required by network coding. Thus, neither Mix-Net nor Onion Routing can be applied to networks equipped with network coding.

Let us examine some typical anonymity techniques designed for wireless networks. Onion Ring [48] is an anonymity scheme proposed for wireless mesh networks. Unfortunately, since Onion Ring is based on Onion Routing, it cannot support network coding either. ANODR [29] provides an untraceable on-demand routing scheme which can protect user identities in multi-hop ad hoc networks. By using broadcast with tap-door information, ANODR supports distributed route discovery between two arbitrary nodes without revealing sender and/or receiver identities. WAR [5] is another anonymity scheme that exploits the broadcast nature of wireless networks. WAR differs from ANODR in that it has the initiating node select the transmission path, and uses cover traffic to thwart global eavesdropping. However, both ANODR and WAR still need to perform layered encryptions/decryptions on packets, just as in Mix-Net. Thus, they still cannot function when wireless network is upgraded to use network coding.

Different from the abovementioned schemes, Crowds [40] is an anonymity scheme designed for web transactions and is not based either in Mix-Net or Onion Routing. In Crowds, each sender will forward its web requests to a set of random chosen members before the request reaches the web server. Thus, Crowds can incur a considerable delay and is not suitable for most applications with requirement of real-time communications. There are some variants of Crowds. For example, Hordes [43] reduces the transmission latency of Crowds using multicast routing, and D-Crowds [13] generalizes Crowds to a TTL-based deterministic forwarding scheme.

References

1. Adeli, M., Liu, H.: Secure network coding with minimum overhead based on hash functions. IEEE Commun. Lett. **13**(12), 956–958 (2009)
2. Agrawal, S., Boneh, D.: Homomorphic macs: MAC-based integrity for network coding. In: Proceedings of Applied Cryptography and Network Security, pp. 292–305 (2009)
3. Benaloh, J.: Dense probabilistic encryption. In: Proceedings of the Workshop on Selected Areas of Cryptography, pp. 120–128 (1994)
4. Bhattad, K., Narayanan, K.R.: Weakly secure network coding. In: Proceedings of International Symposium on Network Coding (NetCod) (2005)
5. Blaze, M., Ioannidis, J., Keromytis, A.D., Malkin, T.G., Rubin, A.: Anonymity in wireless broadcast networks. Int. J. Netw. Secur. **8**(1), 37–51 (2009)
6. Boneh, D., Lynn, B., Shacham, H.: Short signatures from the weil pairing. In: Proceedings of Advances in Cryptology – ASIACRYPT, pp. 514–532 (2001)
7. Boneh, D., Freeman, D., Katz, J., Waters, B.: Signing a linear subspace: signature schemes for network coding. In: Public Key Cryptography (PKC), pp. 68–87. Springer, Heidelberg (2009)
8. Cai, N., Yeung, R.W.: Secure network coding. In: Proceedings of International Symposium on Information Theory (ISIT), p. 323 (2002)
9. Cai, N., Yeung, R.W.: Network error correction, II: lower bounds. Commun. Inf. Syst. **6**(1), 37–54 (2006)
10. Charles, D., Jain, K., Lauter, K.: Signatures for network coding. Int. J. Inf. Coding Theory **1**(1), 3–14 (2009)
11. Chaum, D.L.: Untraceable electronic mail, return addresses, and digital pseudonyms. Commun. ACM **24**(2), 84–90 (1981)
12. Danezis, G., Dingledine, R., Mathewson, N.: Mixminion: design of a type III anonymous remailer protocol. In: Proceedings of IEEE Symposium on Security and Privacy (S&P), pp. 2–15 (2003)
13. Danezis, G., Diaz, C., Käsper, E., Troncoso, C.: The wisdom of crowds: attacks and optimal constructions. In: ESORICS (2009)
14. DiBenedetto, S., Gasti, P., Tsudik, G., Uzun, E.: Andana: anonymous named data networking application. In: Proceedings of the Network and Distributed System Security Symposium (NDSS) (2004)
15. Dingledine, R., Mathewson, N., Syverson, P.: Tor: The second-generation onion router. In: Proceedings of the 13th USENIX Security Symposium (2004)
16. Dong, J., Curtmola, R., Nita-Rotaru, C.: Practical defenses against pollution attacks in intra-flow network coding for wireless mesh networks. In: Proceedings of the Second ACM Conference on Wireless Network Security, pp. 111–122 (2009)
17. El Rouayheb, S.Y., Soljanin, E.: On wiretap networks II. In: Proceedings of International Symposium on Information Theory (ISIT), pp. 551–555 (2007)
18. Fan, Y., Jiang, Y., Zhu, H., Shen, X.: An efficient privacy-preserving scheme against traffic analysis attacks in network coding. In: Proceedings of IEEE INFOCOM, pp. 2213–2221 (2009)

19. Feldman, J., Malkin, T., Stein, C., Servedio, R.: On the capacity of secure network coding. In: Proceedings of the 42rd Allerton Conference on Communication, Control, and Computing (2004)
20. Gkantsidis, C., Rodriguez, P.: Cooperative security for network coding file distribution. In: Proceedings of IEEE INFOCOM, vol. 6, pp. 1–13 (2006)
21. Ho, T., Leong, B., Koetter, R., Médard, M., Effros, M., Karger, D.R.: Byzantine modification detection in multicast networks with random network coding. IEEE Trans. Inf. Theory **54**(6), 2798–2803 (2008)
22. Jacobson, V., Smetters, D.K., Thornton, J.D., Plass, M.F., Briggs, N.H., Braynard, R.L.: Networking named content. In: Proceedings of the 5th International Conference on Emerging Networking Experiments and Technologies (CoNEXT 2009), pp. 1–12 (2009)
23. Jaggi, S., Langberg, M., Katti, S., Ho, T., Katabi, D., Médard, M.: Resilient network coding in the presence of byzantine adversaries. In: Proceedings of IEEE INFOCOM, pp. 616–624 (2007)
24. Jiang, Y., Fan, Y., Shen, X., Lin, C.: A self-adaptive probabilistic packet filtering scheme against entropy attacks in network coding. Comput. Netw. **53**(18), 3089–3101 (2009)
25. Jiang, Y., Zhu, H., Shi, M., Shen, X.S., Lin, C.: An efficient dynamic-identity based signature scheme for secure network coding. Comput. Netw. **54**(1), 28–40 (2010)
26. Katti, S., Cohen, J., Katabi, D.: Information slicing: anonymity using unreliable overlays. In: Proceedings of the 4th USENIX Conference on Networked Systems Design & Implementation (NSDI) (2007)
27. Kehdi, E., Li, B.: Null keys: Limiting malicious attacks via null space properties of network coding. In: Proceedings of IEEE INFOCOM, pp. 1224–1232 (2009)
28. Koetter, R., Kschischang, F.R.: Coding for errors and erasures in random network coding. IEEE Trans. Inf. Theory **54**(8), 3579–3591 (2008)
29. Kong, J., Hong, X.: Anodr: anonymous on demand routing with untraceable routes for mobile ad-hoc networks. In: Proceedings of the 4th ACM International Symposium on Mobile Ad Hoc Networking & Computing, pp. 291–302 (2003)
30. Krohn, M.N., Freedman, M.J., Mazieres, D.: On-the-fly verification of rateless erasure codes for efficient content distribution. In: Proceedings of IEEE Symposium on Security and Privacy (S&P), pp. 226–240 (2004)
31. Le, A., Markopoulou, A.: Tesla-based defense against pollution attacks in p2p systems with network coding. In: Proceedings of International Symposium on Network Coding (NetCod), pp. 1–7 (2011)
32. Li, Q., Chiu, D.M., Lui, J.C.: On the practical and security issues of batch content distribution via network coding. In: Proceedings of IEEE International Conference on Network Protocols (ICNP), pp. 158–167. IEEE, Santa Barbara (2006)
33. Li, Y., Yao, H., Chen, M., Jaggi, S., Rosen, A.: Ripple authentication for network coding. In: Proceedings of IEEE INFOCOM, pp. 1–9 (2010)
34. Lima, L., Médard, M., Barros, J.: Random linear network coding: a free cipher? In: Proceedings of International Symposium on Information Theory (ISIT), pp. 546–550 (2007)
35. Lima, L., Gheorghiu, S., Barros, J., Médard, M., Toledo, A.L.: Secure network coding for multi-resolution wireless video streaming. IEEE J. Sel. Areas Commun. **28**(3), 377–388 (2010)
36. Möller, U., Cottrell, L., Palfrader, P., Sassaman, L.: Mixmaster protocol—version 2. IETF Internet Draft (2003)
37. Ozarow, L., Wyner, A.: Wire-tap channel II. In: Proceedings of Advances in Cryptology, pp. 33–50 (1985)
38. Perrig, A., Canetti, R., Tygar, J.D., Song, D.: Efficient authentication and signing of multicast streams over lossy channels. In: Proceedings of IEEE Symposium on Security and Privacy, pp. 56–73 (2000)
39. Reed, M.G., Syverson, P.F., Goldschlag, D.M.: Anonymous connections and onion routing. IEEE J. Sel. Areas Commun. **16**(4), 482–494 (1998)
40. Reiter, M.K., Rubin, A.D.: Crowds: anonymity for web transactions. ACM Trans. Inf. Syst. Secur. **1**(1), 66–92 (1998)

41. Rennhard, M., Plattner, B.: Introducing morphmix: peer-to-peer based anonymous internet usage with collusion detection. In: Proceedings of ACM Workshop on Privacy in the Electronic Society, pp. 91–102 (2002)
42. Rennhard, M., Plattner, B.: Practical anonymity for the masses with morphmix. In: Financial Cryptography, pp. 233–250. Springer, Heidelberg (2004)
43. Shields, C., Levine, B.N.: A protocol for anonymous communication over the internet. In: Proceedings of the 7th ACM Conference on Computer and Communications Security (CCS), pp. 33–42 (2000)
44. Silva, D., Kschischang, F.R.: Security for wiretap networks via rank-metric codes. In: Proceedings of International Symposium on Information Theory (ISIT), pp. 176–180 (2008)
45. Vilela, J.P., Lima, L., Barros, J.: Lightweight security for network coding. In: Proceedings of IEEE International Conference on Communications, pp. 1750–1754 (2008)
46. Wang, J., Wang, J., Lu, K., Xiao, B., Gu, N.: Optimal linear network coding design for secure unicast with multiple streams. In: Proceedings of IEEE INFOCOM (2010)
47. Wicker, S.B., Bhargava, V.K.: Reed-Solomon Codes and Their Applications. Wiley-IEEE Press, New York (1999)
48. Wu, X., Li, N.: Achieving privacy in mesh networks. In: Proceedings of the Fourth ACM Workshop on Security of Ad Hoc and Sensor Networks, pp. 13–22 (2006)
49. Yeung, R.W., Cai, N.: Network error correction, I: basic concepts and upper bounds. Commun. Inf. Syst. $6(1)$, 19–35 (2006)
50. Yu, Z., Wei, Y., Ramkumar, B., Guan, Y.: An efficient signature-based scheme for securing network coding against pollution attacks. In: Proceedings of IEEE INFOCOM, pp. 1409–1417 (2008)
51. Yu, Z., Wei, Y., Ramkumar, B., Guan, Y.: An efficient scheme for securing xor network coding against pollution attacks. In: Proceedings of IEEE INFOCOM, pp. 406–414 (2009)
52. Yun, A., Cheon, J.H., Kim, Y.: On homomorphic signatures for network coding. IEEE Trans. Comput. $59(9)$, 1295–1296 (2010)
53. Zhang, Z.: Network error correction coding in packetized networks. In: Proceedings of IEEE Information Theory Workshop, pp. 433–437 (2006)
54. Zhao, F., Kalker, T., Médard, M., Han, K.J.: Signatures for content distribution with network coding. In: Proceedings of IEEE International Symposium on Information Theory (ISIT), pp. 556–560 (2007)

Chapter 3
Subspace Authentication for Random Linear Network Coding

As is introduced in Chap. 2, network coding is notoriously susceptible to *pollution attacks*: a single polluted packet can end up corrupting bunches of good ones. The mechanisms that we introduced previously either incur high computation/bandwidth overheads or cannot resist the *tag pollution* proposed recently. This chapter will present a novel idea termed *padding for orthogonality* for network coding authentication. Inspired by it, we design a public key-based signature scheme and a symmetric key-based MAC scheme, which can both effectively contain pollution attacks at forwarders. In particular, we combine them to propose a unified scheme termed *MacSig*, the first hybrid key cryptographic approach to network coding authentication. It can thwart both normal pollution and tag pollution attacks in an efficient way. Simulative results show that our MacSig scheme has a low bandwidth overhead and a verification process 2–4 times faster than typical signature-based solutions in some circumstances.

3.1 Introduction

The information-mixing nature of network coding also renders it more susceptible to *pollution attacks* than traditional store-and-forward paradigm. Consider a scenario in which a commercial data center is distributing a file to a set of costumers via a network-coded P2P network. An adversary pretends as a normal customer, by downloading and contributing packets of the file. In this process, it generates corrupted packets and contributes them to its peers. After being coded with other packets, a single corrupted packet can result in tens or even hundreds of polluted ones. This may cause legitimate users unable to download the file properly.

© Springer International Publishing Switzerland 2016
P. Zhang, C. Lin, *Security in Network Coding*, Wireless Networks,
DOI 10.1007/978-3-319-31083-1_3

Existing schemes include information-theoretic schemes [4, 11, 12, 20, 24] and cryptography-based schemes [1, 3, 6, 8, 9, 13, 15–18, 21–23, 25]. For information-theoretic schemes, they can only passively tolerate pollution at sinks, but not actively prevent them. On the other hand, cryptography-based schemes enable forwarders to verify the integrity of their received packets, so that corrupted packets can be discarded before polluting good ones. This chapter only considers the cryptography-based schemes, which can be further grouped into two classes. The first class includes schemes [3, 6, 9, 13, 16, 17, 21, 23, 25] that are built on public key-based techniques, such as homomorphic hash, homomorphic signature, etc. These schemes are provably secure under the hardness assumptions of well-known cryptographic problems, but will incur high computation overhead at forwarders. The second class are schemes [1, 8, 15, 18, 22] which involve symmetric key encryptions that are computationally efficient. The main disadvantages of schemes in this class include that they incur a larger bandwidth overhead and must carefully manage the keys.

This chapter approaches the problem of network coding authentication using a novel idea called *padding for orthogonality*: the source pads each packet with an extra symbol, so that the subspace spanned by these padded packets is orthogonal to a specific vector; forwarders check the integrity of a received packet by verifying whether it maps this vector to zero. Based on this idea, we propose a public key-based scheme and prove its security under the hardness assumption of discrete logarithm problem. In addition, we also propose a symmetric key-based scheme that is secure against a coalition of c adversaries. Most importantly, we carefully combine them to propose *MacSig*—the first hybrid key-based approach to network coding authentication.

The MacSig scheme offers the following primary features:

- **Security against pollution.** It can effectively not only thwart normal pollution attacks but also resist *tag pollution* presented in [18].
- **Bandwidth efficiency.** It requires a smaller number of tags for each packet compared with [1] (which uses the key distribution scheme given in [5]).
- **Computation efficiency.** It needs a moderate/small number of symmetric key/public key cryptographic operations. Simulations show that its verification process is 2–4 times faster than typical signature-based schemes [3, 21, 25] in some circumstances.

The rest of this chapter is organized as follows. Section 3.2 gives a formal statement of the problem to be studied. Section 3.3 presents our basic idea and introduces two authentication schemes based on it. Section 3.4 proposes a hybrid key authentication scheme, whose performance is evaluated in Sect. 3.5. Section 3.6 discusses how our schemes can be adapted to function in a more general case, and Section 3.7 concludes.

3.2 Problem Statement

3.2.1 Network Model

We consider a typical multicast scenario, in which a source S needs to deliver a series of packets $\underline{x}_1, \underline{x}_2, \ldots, \underline{x}_m$ to multiple receivers $\{R_i\}$. Each packet $\underline{x}_i$ is represented as a vector $(\underline{x}_{i,1}, \underline{x}_{i,2}, \ldots, \underline{x}_{i,n})$ of finite field $\mathbb{F}_p^n$, where p is a prime.

For each $\underline{x}_i$, the source S generates an augmented packet x_i by prefixing $\underline{x}_i$ with the ith unit vector of dimension m:

$$x_i = (\overbrace{0, \cdots, 0, 1, 0, \ldots, 0}^{m}, \underline{x}_{i,1}, \underline{x}_{i,2}, \cdots, \underline{x}_{i,n}) \tag{3.1}$$

$$\underbrace{}_{i-1}$$

Let V denote the subspace spanned by $x_1, x_2, \ldots, x_m$ and term x_i as the ith basis vector of V. Then S sends vectors in V and the network is responsible for replicating of V at each receiver R_i, which can derive $\underline{x}_1, \ldots, \underline{x}_m$ by computing the m basis vectors of V via Gaussian eliminations.

Specifically, for random network coding, the source sends linear combinations of packets using randomly selected coefficients. For example, linearly combining packets $x_1, x_2, \ldots, x_l$ using coefficients $\alpha_1, \alpha_2, \ldots, \alpha_l$ results in

$$y = \sum_i^l \alpha_i x_i = (\sum_i^l \alpha_i x_{i,1}, \ldots, \sum_i^l \alpha_i x_{i,m+n}) \tag{3.2}$$

The first m symbols of y are termed as its *coding coefficients*. Intermediate nodes linearly combines their received packets for output in a similar way. Then a receiver R_i can recover V exactly after receiving m linearly independent packets. In fact, any m received packets are linearly independent with a high probability given the filed size p is sufficiently large [10].

In this chapter, we consider a more realistic setting, in which the data D to be sent consists of more than m packets. Using the technique introduced in [7], S should first break D into multiple *generations*:

$$D = [\underbrace{\underline{x}_1, \ldots, \underline{x}_h}_{G_1}, \cdots, \underbrace{\underline{x}_{(n-1)h+1}, \ldots, \underline{x}_{nh}}_{G_n}, \cdots] \tag{3.3}$$

Then S sends D as a stream of generations, with network coding only performed among packets belonging to the same generation.

3.2.2   Adversary Model

The adversary is aimed at injecting a small number of corrupted packets into the network to cause a large scale of pollution. To achieve this goal, it strives to collect legal packets and forge corrupted ones that can pass the verification of other innocent nodes. Without loss of generality, we assume the source is always trusted, but the relay nodes can be compromised. By compromise, we mean that the adversary can read the memory, monitor the input, and control the output of a compromised node. In this chapter, we allow the adversary to compromise a coalition of nodes to launch more effective attacks. Finally, we assume that the adversaries are aware of our authentication scheme, but are bounded in computation power, and can only perform polynomial-time algorithms.

3.3   Homomorphic Subspace Authentication

In this section, we first introduce the basic idea of *padding for orthogonality*, and then propose two different schemes for network coding authentication: the *homomorphic subspace signature (HSS)* and the *homomorphic subspace MAC (HSM)*.

3.3.1   Basic Idea Overview

Noted from the network model, although packets undergo rounds of coding processes at forwarders, the linear subspace V spanned by them stays constant. We can check the integrity of a packet w by verifying whether $w \in V$. Based on this observation, we can characterize V using a vector v randomly chosen from its orthogonal subspace and let forwarders check whether $w \cdot v^T = 0$. This approach catches an essential property of network coding—the invariance of linear subspace. However, it still has two practical problems unsolved: (1) For different generations, the source should calculate different v's and distribute them prior to transmission, which can cause a high startup latency. (2) These v's should also be authenticated, meaning that an extra secure channel is required.

    Now, we present a novel approach called *padding for orthogonality* to overcome these two problems. By this approach, the source randomly samples a vector $\bar{v}$ of length $m + n + 1$ at the bootstrap stage, and for every generation, the source pads each packet with an extra symbol/tag, so that its inner product with $\bar{v}$ equals zero, as shown in Fig. 3.1. Then the subspace $\bar{V}$ spanned by these augmented packets is orthogonal to $\bar{v}$. To verify a packet w, a relay node just checks whether $w \cdot \bar{v}^T = 0$. Clearly, using this approach, we don't have to pre-distribute $\bar{v}$'s per generation using a secure channel; we just need m tags which cause no startup latency.

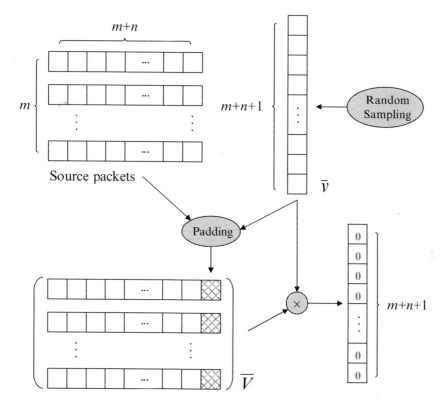

Fig. 3.1 Illustration of the "padding for orthogonality" idea

To make this approach function in the presence of Byzantine adversaries, which attempt to forge tags for illegal packets, we consider the following two techniques. First, we can let the source keep $\bar{v}$ as a secret key and based on it generate a public key which can be used by relay nodes to verify the integrity of packets. If it is sufficiently hard to derive $\bar{v}$ based on the public key, then relay nodes cannot successfully forge tags for any illegal packets. For the second solution, we let the source keep a pool $\mathscr{P}$ of $\bar{v}$'s and pad each packet with multiple tags generated according to these $\bar{v}$'s; each relay node is assigned with a subset of $\mathscr{P}$ and can only verify a packet against part of its tags. If these $\bar{v}$'s are distributed properly, a corrupted packet generated by a malicious node will fail the verifications of other nodes with high probability. The following *homomorphic subspace signature (HSS)* and *homomorphic subspace MAC (HSM)* scheme are designed using these two techniques, respectively. For simplicity of introduction, we assume that the transmission consists of only one generation. We will discuss the multiple-generation cases later in Sect. 3.6.

3.3.2   Homomorphic Subspace Signature

The Model A *Homomorphic Subspace Signature (HSS)* is defined as a tuple of four probabilistic polynomial-time (PPT) algorithms (**Setup**, **Sign**, **Combine**, **Verify**):

- **Setup.** Input: 1^k, the security parameter, and N, the length of vectors to be signed. Output: a prime number q, a secret key K_s, and a public key K_p.
- **Sign.** Input: a vector $x \in \mathbb{F}_q^N$ and the secret key K_s. Output: a vector $\bar{x} = (x, \sigma)$, where $\sigma \in \mathbb{F}_q$ is termed as the signature of x.
- **Combine.** Input: l vectors $\bar{x}_1, \ldots, \bar{x}_l$, where $\bar{x}_i = (x_i \in \mathbb{F}_q^N, \sigma_i \in \mathbb{F}_q)$, and l coefficients $\alpha_1, \ldots, \alpha_l$, where $\alpha_i \in \mathbb{F}_q$. Output: a vector $\bar{x} = (\sum_{i=1}^{l} \alpha_i x_i, \sigma \in \mathbb{F}_q)$.
- **Verify.** Input: a vector $\bar{x} = (x \in \mathbb{F}_q^N, \sigma \in \mathbb{F}_q)$ and the public key K_p. Output: either 1 (accept) or 0 (reject).

An HSS is said to be *correct* if the following two conditions are satisfied:

$$(1)\ \mathbf{Verify}(\mathbf{Sign}(x, K_s), K_p) = 1, \text{ and } (2)\ \mathbf{Verify}(\bar{x}_i, K_p) = 1 \text{ for } i = 1, \ldots, l$$

$$\Rightarrow \mathbf{Verify}(\mathbf{Combine}(\bar{x}_1, \ldots, \bar{x}_l; \alpha_1, \ldots, \alpha_l), K_p) = 1$$

An HSS is said to be *secure* if for any PPT adversary A, the probability that A wins the security game **HSS-GAME** defined below is negligible in the security parameter k:

- **Setup.** The adversary A specifies parameters 1^k and N. The challenger C runs **Setup**$(1^k, N)$ to generate q, K_s, and K_p, of which it sends q and K_p to A.
- **Query.** A adaptively submits vectors $x_1, \ldots, x_m$ to C, which runs **Sign** for these vectors and sends the corresponding $\bar{x}_1, \ldots, \bar{x}_m$ to A.
- **Forge.** A generates a vector $\bar{y} = (y \in \mathbb{F}_q^N, \sigma \in \mathbb{F}_q)$ with $y \notin span(x_1, \ldots, x_m)$. If **Verify**$(\bar{y}, K_p) = 1$, then A wins; otherwise, A loses.

Remarks. When applying the above HSS scheme to the network coding model given in Sect. 3.2.1, we can just let $N = m + n$ and $q = p$.

The Construction Based on the above model, we give our construction of HSS.

- **Setup.** Given 1^k and N, perform the following steps: (1) choose a prime number $q > 2^k$; (2) find a multiplicative cyclic group $\mathbb{G}$ of order q, and select a generator g for $\mathbb{G}$; and (3) set $\boldsymbol{\beta} \xleftarrow{R} \mathbb{F}_q^N \mathbb{F}_q^*$, and calculate $\boldsymbol{h} = (g^{\beta_1}, \ldots, g^{\beta_{N+1}})$. Output q, $K_s = \boldsymbol{\beta}$, and $K_p = \boldsymbol{h}$.
- **Sign.** Given $x \in \mathbb{F}_q^N$ and $\boldsymbol{\beta}$, calculate the signature $\sigma = -(\sum_{i=1}^{N} \beta_i x_i)/\beta_{N+1}$. Output $\bar{x} = (x, \sigma)$.
- **Combine.** Given $\bar{x}_1, \ldots, \bar{x}_l$, where $\bar{x}_i \in \mathbb{F}_q^{N+1}$, and $\alpha_1, \ldots, \alpha_l$, where $\alpha_i \in \mathbb{F}_q$. Output $\bar{x} = \sum_{i=1}^{l} \alpha_i \bar{x}_i$.
- **Verify.** Given $\bar{x} \in \mathbb{F}_q^{N+1}$ and $\boldsymbol{h}$, calculate $\delta = \boldsymbol{h}^{\bar{x}} \triangleq \prod_{i=1}^{N+1} h_i^{\bar{x}_i}$. Output 1 if $\delta = 1$ or 0 otherwise.

> **Theorem 3.1.** *Our construction of HSS is correct.*

Proof. Let $\bar{x}=\mathbf{Sign}(x,\beta)$, then it is easy to verify that $\bar{x} \cdot \beta^T = \sum_{i=1}^{N+1} \bar{x}_i \beta_i = 0$, and then $\delta = h^{\bar{x}} = g^{\bar{x} \cdot \beta^T} = 1$. Thus, $\mathbf{Verify}(\bar{x}, h) = 1$, and Condition (1) holds. Similarly, it is easy to verify that any vector $\bar{x}$ which passes the verification must satisfy $\bar{x} \cdot \beta^T = 0$. Therefore, if we assume $\bar{x}_1, \ldots, \bar{x}_l$ pass the verification, then β is orthogonal to the subspace $\bar{V} = span(\bar{x}_1, \ldots, \bar{x}_l)$. By definition, $\bar{y} = \mathbf{Combine}(\bar{x}_1, \ldots, \bar{x}_l; \alpha_1, \ldots, \alpha_l) \in \bar{V}$, then $\bar{y} \cdot \beta = 0$, and $\mathbf{Verify}(\bar{y}, h) = 1$. Thus, Condition (2) also holds.

> **Theorem 3.2.** *Our construction of HSS is secure.*

Proof. Suppose A wins the security game with some $\bar{y} = (y, \sigma)$. Since $y \notin span(x_1, \ldots, x_m)$, it immediately follows that $\bar{y} \notin span(\bar{x}_1, \ldots, \bar{x}_m)$. In addition, we have $h^{\bar{y}} = 1$ and $h^{\bar{x}_i} = 1$, $i = 1, \ldots, m$. By employing the techniques given in Sect. 3.2 of [2], A can also solve the discrete logarithm problem over $\mathbb{G}$ with a probability at least $1 - 1/q$. Note that for any PPT algorithm, the probability that it solves the discrete logarithm problem over a cyclic group of order $q = 2^k$ is negligible in k [14]. Thus, for any PPT adversary A, the probability that it wins **HSS-GAME** is also negligible in k.

3.3.3 Batched Verification for HSS

From the above construction of HSS, we know that in order to verify a message of $\bar{x}$, a node needs to perform $N + 1$ exponentiation operations on a cyclic group, where N is the length of the message. When the order p of the cyclic group is very large, the computation cost for verification is high. To reduce this cost, we introduce a tree-based batched verification method for HSS. Suppose a node has n messages to be verified, denoted as $\bar{x}_1, \ldots, \bar{x}_n$, then the method works as follows.

First, the node constructs a binary tree T of depth $\lfloor \log_2 n \rfloor + 1$. Each leaf node of T corresponds to a message, and their parent corresponds to a random linear combination of the two messages. For example, in Fig. 3.2, the leaf nodes $< 3, 0 >$ and $< 3, 1 >$ correspond to messages $\bar{x}_1$ and $\bar{x}_2$, respectively. Their parent node $< 2, 0 >$ corresponds to the random linear combination $\alpha_1 \bar{x}_1 + \alpha_2 \bar{x}_2$. To simplify computation, we set the linear coefficients to 1. For example, $\alpha_7, \ldots, \alpha_{10}$ are all set to 1.

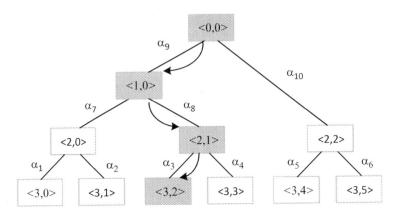

Fig. 3.2 An example of tree-based batched verification for HSS

Note that multiplying each leaf node with a random coefficient is necessary to prevent forgery. To illustrate this, suppose $\bar{x}$ and $\bar{y}$ are legal messages, and the adversary can construct $\bar{x}' = (x_1, \ldots, x_i + a, \ldots, x_N)$ and $\bar{y}' = (y_1, \ldots, y_i - a, \ldots, y_N)$, which can pass the verification. On the other hand, if α_1 and α_2 are randomly selected, then $\alpha_1 \bar{x}' + \alpha_2 \bar{y}'$ will not pass the verification with high probability.

After the tree is constructed, the node begins the verification from the root node $< 0, 0 >$, which corresponds to the message $\sum_{i=1}^n \alpha_i \bar{x}_i$. If this message passes the verification, then the node can conclude that all messages to be verified are legal, and the verification process terminates. Otherwise, there is at least one message that is polluted. In this example, the node would continue to verify $< 1, 0 >$ and $< 2, 2 >$, until the polluted message $\bar{x}_3$ that corresponds to $< 3, 2 >$ is found.

We then give a simple evaluation of this method. If the cost of addition is ignored, then the cost to construct the tree needs $n(N + 1)$ multiplications on the finite field $\mathbb{F}_p$. If there is no polluted message, then only one verification is needed, which costs $N + 1$ exponentiation operations and N multiplications on the cyclic group $\mathbb{G}$. If there is one polluted message, no more than $\lceil \log_2 n \rceil + 1$ verifications are needed to localize that message. On the other hand, n verifications would be always needed without using the batched verification method. Thus, the above tree-based bathed verification method is more cost-effective than straightforward verification approaches.

3.3.4 Homomorphic Subspace MAC

The Model Similar to HSS, a *homomorphic subspace MAC (HSM)* is defined as a tuple of four probabilistic polynomial-time (PPT) algorithms (**Setup, MAC, Combine, Verify**):

- **Setup.** Input: 1^k, the security parameter, and N, the length of vectors to be authenticated. Output: a prime number q, a set K consisting of r MAC keys.
- **MAC.** Input: a vector $x \in \mathbb{F}_q^N$, and the key set K. Output: a vector $\bar{x} = (x, t_1, \ldots, t_r)$, where $t_i \in \mathbb{F}_q$ is a MAC of x calculated using the ith MAC key.
- **Combine.** Input: l vectors $\bar{x}_1, \ldots, \bar{x}_l$, where $\bar{x}_i = (x_i \in \mathbb{F}_q^N, T_i \in \mathbb{F}_q^r)$, and l coefficients $\alpha_1, \ldots, \alpha_l$, where $\alpha_i \in \mathbb{F}_q$. Output: a vector $\bar{x} = (\sum_{i=1}^l \alpha_i x_i, T \in \mathbb{F}_q^r)$.
- **Verify.** Input: a vector $\bar{x} = (x_i \in \mathbb{F}_q^N, T \in \mathbb{F}_q^r)$, and a key set $K' \subset K$. Output: either 1 (accept) or 0 (reject).

An HSM is said to be *correct* if the following two conditions are satisfied:

$$(1)\ \mathbf{Verify}(\mathbf{MAC}(x, K), K) = 1, \text{ and } (2)\ \mathbf{Verify}(\bar{x}_i, K) = 1 \text{ for } i = 1, \ldots, l$$
$$\Rightarrow \mathbf{Verify}(\mathbf{Combine}(\bar{x}_1, \ldots, \bar{x}_l; \alpha_1, \ldots, \alpha_l), K) = 1$$

An HSM is said to be *secure* if for any PPT adversary A, the probability that A wins the security game **HSM-GAME** defined below is no greater than $1/q^d$:

- **Setup.** The adversary A specifies parameters 1^k and N. The challenger C runs **Setup**$(1^k, N)$ to generate q and K. Then it randomly selects two key sets $K' \subset K$ and $K'' \subset K$, with $|K'' \backslash K'| = d$, and sends K' to A.
- **Query.** A adaptively submits vectors $x_1, \ldots, x_m$ to C, which runs **MAC** for these vectors, and sends to A the MACs $T_1, T_2, \ldots, T_m$, where $T_i = \{t_{i,1}, \ldots, t_{i,r}\}$.
- **Forge.** A chooses a vector $y \notin span(x_1, \ldots, x_m)$. Then for each $i = 1, \ldots, r$, it calculates the MAC t_i if $k_i \in K'$ or randomly forges the MAC t_i if $k_i \notin K'$. If **Verify**$((y, t_1, \ldots, t_r), K'') = 1$, then A wins; otherwise, A loses.

Remarks. Different from homomorphic subspace signatures, an HSM uses symmetric keys, i.e., MAC keys, for authentication. The advantage is that forwarders can perform the **Verify** procedure much more efficiently. However, an adversary can also easily forge MACs for illegal packets if all MAC keys are publicized. Thus, we require that the source hold a set K of MAC keys, and each relay node be assigned with a random subset of the K. In this way, each forwarder can only forge some MACs correctly for an illegal packet. If the receiver of this illegal packet has some MAC keys that the adversary does not have, then it can successfully detect the forgery. **HSM-GAME** characterizes this security requirement, by simulating a scenario in which a malicious node with key set K' attempts to forge MACs that pass the verification of another node with key set K''.

The Construction Based on the above model, we give our construction of HSM.

- **Setup.** Given 1^k and N, choose a prime number $q > 2^k$, and set $\gamma_i = (\gamma_{i,1}, \ldots, \gamma_{i,N+1}) \xleftarrow{R} \mathbb{F}_q^N \mathbb{F}_q^*$ for each $i = 1, \ldots, r$. Output $K = (\gamma_1, \ldots, \gamma_r)$.
- **MAC.** Given $x \in \mathbb{F}_q^N$ and K, calculate a tag $t_i = -(\sum_{j=1}^N \gamma_{i,j} x_j)/\gamma_{i,N+1}$ for each $i = 1, \ldots, r$. Output $\bar{x} = (x, t_1, \ldots, t_r)$.

- **Combine.** Given $\bar{x}_1, \ldots, \bar{x}_l$, where $\bar{x}_i \in \mathbb{F}_q^{N+r}$, and $\alpha_1, \ldots, \alpha_l$, where $\alpha_i \in \mathbb{F}_q$, output $\bar{x} = \sum_{i=1}^{l} \alpha_i \bar{x}_i$.
- **Verify.** Given $\bar{x} \in \mathbb{F}_q^{N+r}$ and $K' \subset K$, calculate $\xi_i = \sum_{j=1}^{N} \gamma_{i,j} \bar{x}_j + \gamma_{i,N+1} \bar{x}_{N+i}$, for each $\gamma_i \in K'$. Output 1 if all $\xi_i = 0$ or 0 otherwise.

Theorem 3.3. *Our construction of HSM is correct.*

Proof. For $r = 1$, the proof is much similar to that of Theorem 3.1, and it is easy to extend the proof to cases of $r > 1$. We omit the details here due to the limit of space.

Theorem 3.4. *Our construction of HSM is secure.*

Proof. For each $i = 1, \ldots, r$, we consider the following three cases: (1) $\gamma_i \in K'$. A can accurately calculate the MAC t_i which evaluates ξ_i to zero. (2) $\gamma_i \notin K'$ and $\gamma_i \notin K''$. Any $t_i \in \mathbb{F}_q$ is valid since it will not be checked. (3) $\gamma_i \notin K'$ but $\gamma_i \in K''$. After the **Query** step, A can get the following group of equations regarding γ_i:

$$\begin{pmatrix} x_1, & t_{i,1} \\ x_2, & t_{i,2} \\ \vdots & \vdots \\ x_m, & t_{i,m} \end{pmatrix} \cdot \gamma_i^T = 0 \tag{3.4}$$

which has p^{N+1-R} solutions for γ_i, where R is row rank of the coefficient matrix. Suppose we insert into this group of equations with

$$(y, t) \cdot \gamma_i^T = 0 \tag{3.5}$$

where $y \notin span(x_1, \ldots, x_m)$, $t \in \mathbb{F}_q$. Then the row rank of the coefficient matrix will be $R + 1$; the solution set for γ_i will have cardinality q^{N-R}. This means that $q^{N-R}/q^{N+1-R} = 1/q$ of solutions to Eq. (3.4) can solve Eq. (3.5). As we assume that the MAC key γ_i is sampled randomly from $\mathbb{F}_q^{N+1}$, the probability any t is a valid MAC that evaluates ξ_i to zero is exactly $1/q$. Since there are totally d such i satisfying that $\gamma_i \notin K'$ and $\gamma_i \in K''$, the probability of $\xi_i = 0$ for all these i is then $1/q^d$.

3.3.5 Key Distribution for HSM

Recall in our construction of HSM, the probability that any packet polluted by a node A can pass the verification of another node B is bounded by $1/q^d$, where d is the number of MAC keys held by B but not by A. This implies that the security level of HSM depends on how MAC keys are distributed among nodes in the network. In this subsection, we first formalize this problem of MAC key distribution and then introduce our proposed scheme. We assume a strong adversary model, in which a set of compromised nodes can collude to launch pollution attacks.

The Problem Let Ω denote the set of all nodes except the source S, with $|\Omega| = N$. Let K be the set of all MAC keys held by S. To each node $\omega_i \in \Omega$, S assigns a subset $K(\omega_i) \subset K$ of MAC keys. For a set $A \subset \Omega$, define its keys as $K(A) \triangleq \bigcup_{\omega_i \in A} K(\omega_i)$. We say a key $k \in K$ is *safe* with respect to a node ω_i and a set A, if $k \in K(\omega_i) \backslash K(A)$. We say the key distribution scheme is *c-secure* if for any $\omega_i \in \Omega$ and $A \subset \Omega$ with $|A| \le c$, there is at least one safe key.

Canetti et al. [5] introduce a probabilistic key distribution mechanism, in which every node $\omega_i \in \Omega$ is assigned with any key $k \in K$ with an equal probability P_a. They show that by letting $|K| = e(c + 1) \ln \frac{1}{\epsilon}$ and $P_a = \frac{1}{c+1}$, the probability that there is at least one safe key for a randomly chosen ω_i and A with $|A| = c$ can be made higher than $1 - \epsilon$. However, to achieve this for any ω_i and A with $|A| = c$, i.e., to make the distribution mechanism c-secure with probability at least $1 - \epsilon$, $|K|$ should be made no less than $e(c + 1)^2 \ln N$. This means that each packet should carry $e(c + 1)^2 \ln N$ MACs, which clearly does not scale when the network size N is large. To overcome this limitation, we propose a new approach using which the number of MACs per packet has no relation with N.

Double-Random Key Distribution Our proposed scheme, termed as *double-random key distribution*, gets its name because MAC keys are distributed via two random procedures: the first procedure assigns each node with a random set of keys, just like in [5]; the second one randomly selects keys to be used for MAC calculations. More specifically, in the second procedure, the source randomly selects a subset of l MAC keys from K for each generation. Then the source calculates l MACs using these keys for each packet. In addition, to inform forwarders of the selected keys, the source attaches the indexes of these l keys to each packet.

The rationale of our proposed scheme is to introduce randomness when generating MACs at the source. This randomness can prevent the adversary from knowing the keys used for MAC calculation beforehand and hence prevent the adversary from electively compromising nodes. Theorem 3.5 shows that the number of MACs per packet has no relation with the N, meaning that the bandwidth overhead is scalable with the network size.

Theorem 3.5. *Let the number of secret keys be* $|K| = e(c+1)m$, *with* $m = \frac{2}{\delta^2}(\gamma+1)(c+1)\ln N$, $\gamma > 0$ *and* $0 < \delta < 1$. *Let the number of MACs per packet be* $l = \frac{1}{1-\delta}e(c+1)\ln\frac{1}{\epsilon}$ *and the key assigning probability be* $P_a = \frac{1}{c+1}$. *Then the probability that the double-random key distribution is c-secure is no smaller than* $1 - \epsilon$, *when* $\gamma \to \infty$.

Proof. For each node ω_i and any set A of c nodes, the probability that a specific key k is safe can be calculated by

$$\Pr(k \text{ is safe}|\omega_i, A) = (1 - \frac{1}{c+1})^c \frac{1}{c+1} < \frac{1}{e(c+1)} \tag{3.6}$$

Then the expected number of safe keys can be derived by

$$E(\# \text{ of safe keys}|\omega_i, A) = \frac{|K|}{e(c+1)} = m \tag{3.7}$$

Using Chernoff bound, we have

$$\Pr(\# \text{ of safe keys} < (1-\delta)m|\omega_i, A) < \exp(-\frac{\delta^2}{2}m) \tag{3.8}$$

Based on this, the probability that there exist a node ω_i and a set A of c nodes, such that the number of safe keys is less than $(1-\delta)m$, can be calculated by

$$\begin{aligned}
h &= \Pr(\bigcup_{A,\omega_i}(\# \text{ of safe keys} < (1-\delta)m)|\omega_i, A) \\
&< \sum_{\omega_i, A} \Pr(\# \text{ of safe keys} < (1-\delta)m|\omega_i, A) \\
&< n\binom{N}{c}\exp(-\frac{\delta^2}{2}m) < N^{c+1}\exp(-\frac{\delta^2}{2}m) \\
&= \exp((c+1)\ln N - \frac{\delta^2}{2}\frac{2}{\delta^2}(\gamma+1)(c+1)\ln N)) \\
&= \exp(-\gamma(c+1)\ln N) \\
&= N^{-\gamma(c+1)}
\end{aligned}$$

As $\gamma \to \infty$, $h \to 0$, meaning that the number of safe keys given any ω_i and A is no less than $(1-\delta)m$. Then the probability that a randomly chosen key from K is safe for all A and ω_i is

$$r > \frac{(1-\delta)m}{|K|} = \frac{1-\delta}{e(c+1)} \tag{3.9}$$

If the source randomly chooses l keys from K, then the probability that no safe key is selected can be calculated by

$$u = (1 - r)^l < (1 - \frac{1 - \delta}{e(c + 1)})^{\frac{1}{1-\delta}e(c+1)\ln\frac{1}{\epsilon}} < \epsilon. \tag{3.10}$$

Thus, the double-random key distribution scheme is c-secure with probability no less than $1 - \epsilon$.

3.4 The MacSig Authentication Scheme

As shown in the previous section, the HSS scheme is proven secure under the hardness assumption of discrete logarithm problem, and it incurs a lower bandwidth overhead than HSM. However, in most cases, especially when computation power is constrained (e.g., WSN), we prefer the HSM scheme, since it puts less burden on forwarders. However, the recent work [18] reports that such homomorphic MAC-based schemes (including our HSM) may suffer from *tag pollution*. Fortunately, we observe that our HSS scheme can be utilized to help HSM thwart this attack. In this section, we first give preliminary to the problem of tag pollution and then propose a novel scheme termed *MacSig* to solve it.

3.4.1 *Tag Pollution*

By tag pollution, an adversary aims to modify the tags (MACs for HSM) carried by packets rather than the contents of them. If a receiver of a packet with polluted tags does not have necessary keys to check at least one of them, it cannot detect and filter out this tag-polluted packet. It is possible that a packet with polluted tags travels multiple hops until it is finally detected and discarded, which can result in a waste of network bandwidth.

For a concrete understanding, consider the example given by Fig. 3.3, in which there are one adversary A, one receiver D, and some relay routers. Each packet is attached with four MACs. A pollutes two MACs t_1 and t_3 in both its output packets x_1 and x_2. As R_1 can verify the packet x_1 against its MAC t_3 using k_3, it can detect the tag pollution and discard x_1. However, since R_2 has neither k_1 nor k_3, x_2 will pass the verification of R_2 and encode into another two packets x_3 and x_4. As D just extracts the content of x_3, it is not affected by the polluted MACs. On the other hand, packet x_4 will be discarded at node R_3. As a result, the bandwidth of link $R_2 \rightarrow R_3$ is wasted. For a worse case in which a tag-polluted packet can travel more hops and infect more packets before being detected, the bandwidth waste will be considerable.

To thwart tag pollution, Li et al. [18] propose to delay the release of MAC keys until they are to be used. Though being very effective, this solution requires that

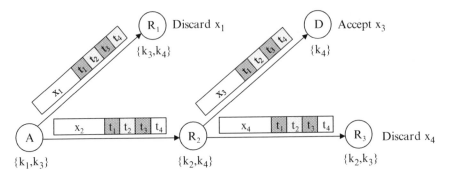

Fig. 3.3 An example of tag pollution in HSM

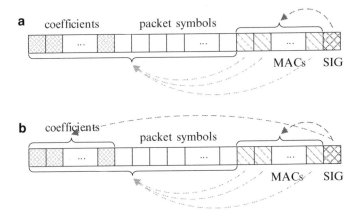

Fig. 3.4 Basic idea of the MacSig authentication scheme

nodes be globally synchronized, and the source needs a time-based scheduler to distribute keys. These requirements are difficult to meet in real distributed scenarios, e.g., communication networks.

3.4.2 MacSig: Homomorphic MAC + Signature

In this chapter, we introduce a novel scheme termed *MacSig*, which uses both homomorphic MACs and signatures for packet authentication. The basic idea is shown in Fig. 3.4a, where the packet content is authenticated by homomorphic MACs, and these MACs are further authenticated by a homomorphic signature. In real implementation of MacSig, we also let the signature authenticate part of the packet content (reasons to be given in Sect. 3.4.3). Specifically, we let the signature authenticate both the MACs and coding coefficients of the packet, as shown in Fig. 3.4b.

In the following, we give the details of our MacSig scheme. For convenience of demonstration, we first assume the key distribution scheme in [5] and then show how our double-random approach can be used in MacSig. Similar to the construction of HSS and HSM introduced in Sect. 3.3, our MacSig consists of four probabilistic polynomial-time (PPT) algorithms.

Setup The source performs the following five steps: (1) Find a multiplicative cyclic group $\mathbb{G}$ of order p, and select a generator g for $\mathbb{G}$. (2) Sample the secret key $\boldsymbol{\beta} = (\beta_1, \ldots, \beta_{m+l+1}) \xleftarrow{R} \mathbb{F}_p^{m+l}\mathbb{F}_p^*$. (3) Compute the public key $\boldsymbol{h} = g^{\boldsymbol{\beta}} \triangleq (g^{\beta_1}, \ldots, g^{\beta_{m+l+1}})$. (4) Sample a pool of MAC keys $K = \{\boldsymbol{\gamma}_i\}_{i=1}^l$, where $\boldsymbol{\gamma}_i = (\gamma_{i,1}, \ldots, \gamma_{i,m+n+1}) \xleftarrow{R} \mathbb{F}_p^{m+n}\mathbb{F}_p^*$. (5) For each MAC key, assign it to each node with an equal probability P_a.

MAC+Sign The source performs the following three steps: (1) For the ith source packet x_i, attach l MACs $\{t_{i,1}, \ldots, t_{i,l}\}$, where $t_{i,j}$ is calculated as

$$t_{i,j} = \frac{-\sum_{r=1}^{m+n} \gamma_{j,r} x_{i,r}}{\gamma_{j,m+n+1}} \tag{3.11}$$

(2) Attach the signature σ_i to x_i, where σ_i is calculated as

$$\sigma_i = -\frac{\sum_{j=1}^m \beta_j x_{i,j} + \sum_{j=1}^l \beta_{j+m} t_{i,j}}{\beta_{m+l+1}} \tag{3.12}$$

(3) Output the resultant packet $\bar{x}_i = (x_i, t_{i,1}, \ldots, t_{i,l}, \sigma_i)$, which has length $m + n + l + 1$.

Combine Relay nodes perform standard random network coding over incoming packets to generate output ones.

Verify For each incoming packet $\bar{y}$, the relay node calculates the value of δ using

$$\delta = \prod_{i=1}^m h_i^{\bar{y}_i} \prod_{i=1}^{l+1} h_{m+i}^{\bar{y}_{m+n+i}} \tag{3.13}$$

$\bar{y}$ is rejected if $\delta \neq 1$; otherwise, the verification process continues. Using each of its MAC keys $\bar{\gamma}_i$, the relay node calculates the value of ξ_i using

$$\xi_i = \sum_{r=1}^{m+n} \gamma_{i,r} \bar{y}_r + \gamma_{i,m+n+1} \bar{y}_{m+n+j} \tag{3.14}$$

$\bar{y}$ is rejected if these exists a $\xi_i \neq 0$; otherwise, it is accepted.

To employ the double-random key distribution in MacSig, we alternatively let the source sample $|K| > l$ MAC keys and randomly select l from them to calculate MACs in each generation. Also, the source should attach the indexes of the l selected keys to each packet $\bar{x}_i$. The value of $|K|$ and l can be determined according to Theorem 3.5.

3.4.3   Security Analysis of MacSig

Theorem 3.6. *The MacSig authentication scheme is secure against tag pollution.*

Proof. First, we give some notations in this proof. Let $\bar{x} = (x, t)$ denote a packet, where $x = (x_1, x_2, \ldots, x_m)$ are message symbols, and $t = (t_1, t_2, \ldots, t_l, \sigma)$ are tags. Let $\bar{V}$ be the subspace spanned by the m source packets $\bar{x}_1, \bar{x}_2, \ldots, \bar{x}_m$ and T be the subspace spanned by their respective tags $t_1, t_2, \ldots, t_m$.

We consider an attack scenario comprised of an adversary node A and an innocent node I. Let K_I be the set of MAC keys assigned to I, with $b = |K_I| > 0$. Without loss of generality, we assume $K_I = \{\gamma_1, \gamma_2, \ldots, \gamma_b\}$, with γ_i corresponding to the ith MAC of a packet. The goal of A is defined as: given a packet $\bar{y} = (y, t) \in \bar{V}$, construct tags $t' \neq t$ so that the packet $\bar{y}' = (y, t')$ passes I's verification.

Suppose that the authentication approach given in Fig. 3.4a is employed. Then t' should satisfy the following two conditions: (1) $t' \in T$ (by Theorem 3.2); (2) $t'_i = t_i$ for each $i = 1, 2, \ldots, b$ (since message symbols are not modified). Let $\bar{y}_1, \bar{y}_2, \ldots, \bar{y}_m$ be any m linearly independent packets belonging to $\bar{V}$, with $\bar{y}_i = (y_i, t_i)$. Then any t' that satisfies the above two conditions can be represented as

$$t' = \alpha (t_1^T, t_2^T, \ldots, t_m^T)^T \tag{3.15}$$

with α subjected to

$$\alpha \cdot \begin{pmatrix} t_{1,1}, & t_{1,2}, & \cdots & t_{1,b} \\ t_{2,1}, & t_{2,2}, & \cdots & t_{2,b} \\ \vdots & \vdots & \ddots & \vdots \\ t_{m,1}, & t_{m,2}, & \cdots & t_{m,b} \end{pmatrix} = (t_1, t_2, \ldots, t_b) \tag{3.16}$$

Let λ be the column rank of the coefficient matrix. Assume that $b < m$, then we can obtain $p^{m-\lambda}$ solutions for α, meaning that there will be $p^{m-\lambda}$ possible t'. As only one of these t' is equal to t, the tag pollution will succeed with a probability of $1 - 1/p^{m-\lambda}$. Thus, the verification approach given by Fig. 3.4a is not secure.

Now suppose the MacSig scheme is employed instead. In this case, Condition (2) remains the same, while Condition (1) should be changed to $(y_1, y_2, \ldots, y_m, t') \in T'$, where T' is the subspace spanned by the first m and the last $l + 1$ symbols of source packets $\bar{x}_1, \bar{x}_2, \ldots, \bar{x}_m$. As a result, the equations that α is subjected to become

$$\alpha \cdot \begin{pmatrix} y_{1,1}, & \ldots, & y_{1,m}, & t_{1,1}, & \ldots & t_{1,b} \\ y_{2,1}, & \ldots, & y_{2,m}, & t_{2,1}, & \ldots & t_{2,b} \\ \vdots & \ddots & \vdots & \vdots & \ddots & \vdots \\ y_{m,1}, & \ldots, & y_{m,m}, & t_{m,1}, & \ldots & t_{m,b} \end{pmatrix} = (y_1, \ldots, y_m, t_1, \ldots, t_b) \qquad (3.17)$$

where the first m columns are introduced by the new Condition (1). As these m columns are linearly independent, the column rank of the coefficient matrix will be m. Then there is only one solution for α, using which A can only construct a $t' = t$. Thus, tag pollution is not possible.

3.5 Performance Evaluation

This section studies the performance of our MacSig authentication scheme in terms of bandwidth and computation overhead.

3.5.1 Bandwidth Overhead

We neglect the bandwidth consumed by the key distribution in the **Setup** stage, since it can be done offline. The online bandwidth overhead per packet includes l MACs, l MAC key indexes, and a signature. Since MACs and signatures are defined over $\mathbb{F}_p$, just as message symbols, they both have size of $|p| = \lceil \log_2 p \rceil$ bits. Next, we determine the size of a MAC key index. As there are $|K| = \frac{2}{\delta^2} e(c+1)^2 (\gamma + 1) \ln N$ MAC keys in total, a key index can be represented by $\lambda = \lceil \log_2 |K| \rceil$ bits. To get a sufficiently large λ, we assume an extreme case where $c = 10$, $\gamma = 1000$, and $N = 1000$ (then the probability h in the proof of Theorem 3.5 will be less than 10^{-12}). For this case, λ equals 29, but we choose $\lambda = 32$ instead for the sake of byte alignment.

From the above analysis, the bandwidth overhead of the MacSig scheme is $O_b = \frac{l+1}{m+n} + \frac{32l}{|p|(m+n)}$, where $l = \frac{1}{1-\delta} e(c+1) \ln \frac{1}{\epsilon}$. We calculate O_b by fixing $\delta = 0.1$, $n = 20m$, $|p| = 128$ and varying the values of n, c, ϵ. Figure 3.5 shows the relationship between bandwidth overhead per packet and packet size n (the number of symbols in each packet) for different c and ϵ. We observe that the bandwidth overhead decreases with the packet size. Especially, when the packet size is larger than 700 symbols and the number of colluding adversaries is less than 3, the per-packet bandwidth overhead sits between 5 % and 10 %.

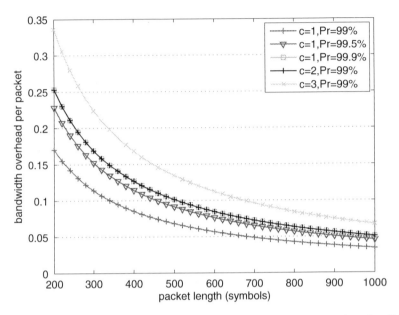

Fig. 3.5 The per-packet bandwidth overhead of our MacSig scheme (c is the number of coalition, and $Pr = 1 - \epsilon$ is the security probability)

3.5.2 Computation Overhead

For a similar reason with the last subsection, we will only consider the online computation overhead incurred by the last three stages.

- **MAC+Sign**. For each packet, the source needs $(m + n + 1)l$ multiplications to calculate l MACs and $m + l + 1$ multiplications to generate the signature for these MACs. Thus, the computation overhead in this stage is $(m+n+1)l+(m+l+1)$ multiplications per packet.

- **Combine**. Let w be the average number of packets combined in each network coding round, then an average of $w(l + 1)$ multiplications are needed to combine the MACs and signatures of the corresponding packets. The computation overhead in this stage will not burden the relay nodes much more, considering the standard network coding still requires $w(m + n)$ multiplications, with the packet size $m + n$ much larger than the number of MACs l.

- **Verify**. First of all, we need to determine the computation complexity of exponentiation over $\mathbb{F}_p$, since it is a key operation required by MacSig verification. We utilize the typical "square and multiple" method to calculate y^x over $\mathbb{F}_p$ as follows. First we compute the value of y^{2^z} for $1 \le z < |x|$, where $|x|$ is the size of x in bits. Since half of the bits of x are zero on average, we need another $\frac{1}{2} \log_2 |x|$ multiplications to obtain y^x. Thus, an exponentiation over $\mathbb{F}_p$ takes $\frac{3}{2}|p|$ multiplications. Secondly, to obtain a benchmark for the running time of multiplications over $\mathbb{F}_p$, we implement the Montgomery multiplication algorithm

[19] on a 2.00 GHz Intel Core 2 CPU. For the case of $|p| = 128$, we observe that roughly 2.5×10^5 multiplications over $\mathbb{F}_p$ can be performed per second. Recall that to verify a packet in MacSig, $m + l + 1$ exponentiations and $(m + n + 1)l$ multiplications are needed. Using the fact that an exponentiation is equivalent to $\frac{3}{2}|p|$ multiplications, the overhead of this stage is $\frac{3}{2}|p|(m+l+1) + (m+n+1)l$ multiplications. Then we use the benchmark developed above to evaluate the verification time of our MacSig scheme. By fixing $\delta = 0.1, n = 20m, |p| = 128$, $\epsilon = 1/100$, and varying n, c, we obtain the results as shown in Fig. 3.6. For comparison, we also include the other three signature-based schemes [3, 21, 25], all of which require no less than $m + n$ exponentiations to verify a packet. From Fig. 3.6, we observe that the verification process in our MacSig scheme is 2–4 times faster than those of the other three.

3.6 Discussion

In the above of this chapter, we have introduced three authentication schemes (HSS, HSM, and MacSig) by assuming single generation. If the transmission consists of multiple generations, the adversary can launch repetitive attack: collect legitimate

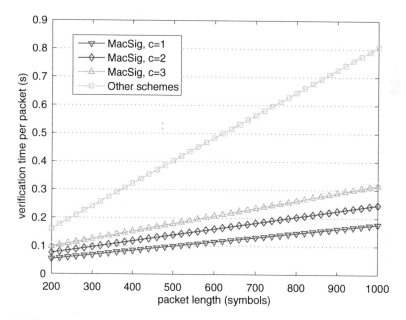

Fig. 3.6 The per-packet verification time of our MacSig scheme and those of the other three schemes [3, 21, 25] (represented as a *single curve*)

packets of previous generations and use them to fake packets for subsequent ones. In the following, we discuss how to improve HSM and HSS so that they can resist this attack.

For the HSM scheme, we assume that the source and other nodes share a common pseudorandom number generator (PRNG). At the Setup stage, the source samples a pool of random seeds and assigns these seeds to other nodes the same way as distributing MAC keys. Before each generation, all nodes run PRGN using their seeds to generating new MAC keys. In this way, different generations will use different MAC keys, and thus repetitive attack is not possible.

For the HSS scheme, we consider the following two approaches. In the first approach, prior to each generation, the source randomly alters one element in the secret key and informs other nodes of the changed element in the public key, just like the scheme in [25]. This approach can be effective when the adversary is unable collect legitimate packets with many zeros.

In the second approach, we let the signature of a source packet x_i be $\sigma_i = g^{-\sum_{j=m+1}^{m+n} \beta_j x_{i,j}}$. For each generation, the source signs and sends the m signatures $\{\sigma_i\}_{i=1}^{m}$ of its packets, and forwarders check whether $\prod_{i=m+1}^{m+n} h_i^{y_i} \prod_{i=1}^{m} \sigma_i^{y_i} = 1$. Alternatively, since this signature scheme is homomorphic, we can have each signature travel with the packet it signs. In this way, we don't need an extra channel to transmit signatures in advance.

3.7 Conclusion

This chapter studies the problem of network coding authentication in the presence of pollution attacks and tag pollution attacks. The basic idea is to pad each source packet with an extra symbol to make it orthogonal to a given vector. Under this idea, we propose a signature-based scheme HSS. HSS doesn't need to pre-distribute signatures for each generation and hence incurs no startup latency. We also propose a MAC-based scheme HSM, which employs a double-random key distribution approach. The main advantages of HSM is that the number of MACs attached to each packet scales with the network size. By utilizing the techniques of both HSS and HSM, we finally propose a hybrid key-based authentication scheme MacSig. We demonstrate that the MacSig scheme can effectively resist both normal pollution and tag pollution attacks, while incurring a relatively low overhead in bandwidth and computation.

References

1. Agrawal, S., Boneh, D.: Homomorphic macs: MAC-based integrity for network coding. In: Proceedings of Applied Cryptography and Network Security, pp. 292–305 (2009)
2. Boneh, D., Franklin, M.: An efficient public key traitor tracing scheme. In: Advances in Cryptology, pp. 783–783. Springer, Heidelberg (1999)

3. Boneh, D., Freeman, D., Katz, J., Waters, B.: Signing a linear subspace: signature schemes for network coding. In: Public Key Cryptography (PKC), pp. 68–87. Springer, Heidelberg (2009)
4. Cai, N., Yeung, R.W.: Network error correction, II: lower bounds. Commun. Inf. Syst. **6**(1), 37–54 (2006)
5. Canetti, R., Garay, J., Itkis, G., Micciancio, D., Naor, M., Pinkas, B.: Multicast security: a taxonomy and some efficient constructions. In: Proceedings of IEEE INFOCOM, pp. 708–716 (1999)
6. Charles, D., Jain, K., Lauter, K.: Signatures for network coding. Int. J. Inf. Coding Theory **1**(1), 3–14 (2009)
7. Chou, P.A., Wu, Y., Jain, K.: Practical network coding. In: Proceedings of Annual Allerton Conference on Communication Control and Computing, pp. 40–49 (2003)
8. Dong, J., Curtmola, R., Nita-Rotaru, C.: Practical defenses against pollution attacks in intra-flow network coding for wireless mesh networks. In: Proceedings of the second ACM Conference on Wireless Network Security, pp. 111–122 (2009)
9. Gkantsidis, C., Rodriguez, P.: Cooperative security for network coding file distribution. In: Proceedings of IEEE INFOCOM, vol. 6, pp. 1–13 (2006)
10. Ho, T., Médard, M., Koetter, R., Karger, D.R., Effros, M., Shi, J., Leong, B.: A random linear network coding approach to multicast. IEEE Trans. Inf. Theory **52**(10), 4413–4430 (2006)
11. Ho, T., Leong, B., Koetter, R., Médard, M., Effros, M., Karger, D.R.: Byzantine modification detection in multicast networks with random network coding. IEEE Trans. Inf. Theory **54**(6), 2798–2803 (2008)
12. Jaggi, S., Langberg, M., Katti, S., Ho, T., Katabi, D., Médard, M.: Resilient network coding in the presence of byzantine adversaries. In: Proceedings of IEEE INFOCOM, pp. 616–624 (2007)
13. Jiang, Y., Zhu, H., Shi, M., Shen, X.S., Lin, C.: An efficient dynamic-identity based signature scheme for secure network coding. Comput. Netw. **54**(1), 28–40 (2010)
14. Katz, J., Lindell, Y.: Introduction to Modern Cryptography: Principles and Protocols. Chapman & Hall/CRC, Boca Raton (2007)
15. Kehdi, E., Li, B.: Null keys: Limiting malicious attacks via null space properties of network coding. In: Proceedings of IEEE INFOCOM, pp. 1224–1232 (2009)
16. Krohn, M.N., Freedman, M.J., Mazieres, D.: On-the-fly verification of rateless erasure codes for efficient content distribution. In: Proceedings of IEEE Symposium on Security and Privacy (S&P), pp. 226–240 (2004)
17. Li, Q., Chiu, D.M., Lui, J.C.: On the practical and security issues of batch content distribution via network coding. In: Proceedings of IEEE International Conference on Network Protocols (ICNP), pp. 158–167. IEEE, Santa Barbara (2006)
18. Li, Y., Yao, H., Chen, M., Jaggi, S., Rosen, A.: Ripple authentication for network coding. In: Proceedings of IEEE INFOCOM, pp. 1–9 (2010)
19. Montgomery, P.L.: Modular multiplication without trial division. Math. Comput. **44**(170), 519–521 (1985)
20. Yeung, R.W., Cai, N.: Network error correction, I: basic concepts and upper bounds. Commun. Inf. Syst. **6**(1), 19–35 (2006)
21. Yu, Z., Wei, Y., Ramkumar, B., Guan, Y.: An efficient signature-based scheme for securing network coding against pollution attacks. In: Proceedings of IEEE INFOCOM, pp. 1409–1417 (2008)
22. Yu, Z., Wei, Y., Ramkumar, B., Guan, Y.: An efficient scheme for securing xor network coding against pollution attacks. In: Proceedings of IEEE INFOCOM, pp. 406–414 (2009)
23. Yun, A., Cheon, J.H., Kim, Y.: On homomorphic signatures for network coding. IEEE Trans. Comput. **59**(9), 1295–1296 (2010)
24. Zhang, Z.: Network error correction coding in packetized networks. In: Proceedings of IEEE Information Theory Workshop, pp. 433–437 (2006)
25. Zhao, F., Kalker, T., Médard, M., Han, K.J.: Signatures for content distribution with network coding. In: Proceedings of IEEE International Symposium on Information Theory (ISIT), pp. 556–560 (2007)

Chapter 4
Lightweight Encryption for Random Linear Network Coding

Researchers have shown that network coding can help reduce the energy consumption of communication by reducing transmissions. However, apart from transmission cost, there are other sources of energy consumption, e.g., data encryption/decryption. This chapter will study how to leverage network coding to reduce the energy consumed by data encryption. It is interesting that network coding has a nice property of intrinsic security, based on which encryption can be done quite efficiently. To this end, this chapter presents P-Coding, a lightweight encryption scheme to provide confidentiality for network coding. The basic idea of P-Coding is to let the source randomly permutes the symbols of each packet (which is prefixed with its coding vector), before performing network coding operations. Without knowing the permutation, eavesdroppers cannot locate coding vectors for correct decoding and thus cannot obtain any meaningful information. The network setting is mobile ad hoc networks (MANETs), where energy saving is a very important research issue. We demonstrate that due to its lightweight nature, P-Coding incurs minimal energy consumption compared to other encryption schemes in MANETs.

4.1 Introduction

A lot of research efforts have focused on how to reduce energy consumption in mobile ad hoc networks (MANETs) [6, 25, 30]. Recent studies demonstrate that network coding can help achieve a lower energy consumption in MANETs [10, 16, 31]. The energy saving comes from the fact that less transmissions are required when in-network nodes are enabled to encode packets. The basic idea can be illustrated using the following example. Suppose there are six nodes forming a hexagon, and the transmission range of each node can only reach its left and right neighbor. Each node needs to broadcast one message to all other

© Springer International Publishing Switzerland 2016
P. Zhang, C. Lin, *Security in Network Coding*, Wireless Networks,
DOI 10.1007/978-3-319-31083-1_4

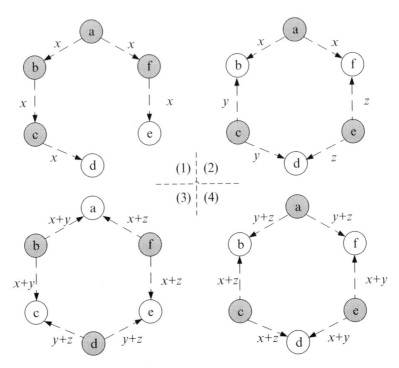

Fig. 4.1 Example illustrating how network coding reduces transmission times in MANETs. The *shaded nodes* are those involved in transmissions

nodes. Without network coding, each message would require four broadcasts, as shown in Fig. 4.1(1). With network coding (Fig. 4.1(2)–(4)), a total number of nine transmissions are needed for three messages, i.e., three transmissions per message. If we would not consider the energy consumed by encoding and decoding operations, this means one-fourth energy can be saved.

Besides basic transmissions, energy consumption can also come from encryption and decryption operations at each node, as most MANETs need some level of protection on their content. For example, in a battlefield, the data communicated between soldiers with mobile devices can be very sensitive and should be kept confidential during transmissions. The straightforward approach to provide confidentiality for network-coded MANETs is to encrypt the packet payload using symmetric key encryption algorithms. While this method is not that efficient, reference [21] shows that on a Motorola's "DragonBall" embedded microprocessor, it consumes around $13.9\,\mu J$ to send a bit, while consumes another $7.9\,\mu J$ per bit when symmetric key algorithms are used. In fact, the information mixing feature of network coding provides an intrinsic security, based on which a more efficient cryptographic scheme can be designed. Vilela et al. [17, 27] propose such a scheme, in which the source performs random linear coding on the messages to be sent and locks/encrypts the coding vectors using the symmetric key shared between it and all sinks. Fan et al. [9] propose to encrypt coding vectors using homomorphic encryption functions (HEFs) in an end-to-end manner. Due to the homomorphic nature of HEFs, network coding can be performed directly on the encrypted coding vectors, without impacting the standard network coding operations. However, the above two approaches have large overhead with respect to either computation or space and may not be suitable for MANETs.

This chapter will present a new encryption scheme that can fully exploit the security property of network coding. The key observation underlying this scheme is: *since both the coding vectors and message content are necessary for decoding, randomly reordering/mixing them will generate considerable confusion to the eavesdropping adversary.* The encryption scheme, named P-Coding, is a lightweight encryption scheme that can fight against eavesdroppers in network-coded MANETs. In a nutshell, P-Coding randomly mixes symbols of each coded packet (packet prefixed with its coding vector) using *permutation encryption*, to make it hard for eavesdroppers to locate coding vectors for packet decoding.

The contribution of this chapter is twofold: (1) we propose a new encryption scheme which is lightweight in computation by leveraging network coding and thus very attractive in network-coded MANETs to further reduce energy consumption, and (2) we present an analysis on the intrinsic weak security provided by network coding, which is more accurate than [2]. We show that network coding is inherently weakly secure with high probability, when the coding vectors are randomly chosen over a large finite field.

The remainder of this chapter is organized as follows. Section 4.2 presents the system model and security model. Section 4.3 evaluates the intrinsic security provided by network coding. Section 4.4 introduces the P-Coding scheme and its enhanced version, and their security is analyzed in Sect. 4.5. Section 4.6 evaluates the performance of P-Coding with analysis and experiments. Section 4.7 discusses remaining issues, and Section 4.8 concludes.

4.2 Problem Statement

4.2.1 Network Model

We consider a typical MANET consisting of N nodes, each of which can be a source. The MANET can be modeled as an acyclic directed graph $G = (V, E)$. For each node $v \in V$, there is a link from v to u if u is within v's transmission range. Let $\Gamma^-(v)$ be the set of links terminating at v and $\Gamma^+(v)$ be the set of links originating from v. We assume that each link $e \in E$ has the capacity of one packet per unit time, and $y(e)$ is the packet carried on it. Here a packet is defined as a row vector of l elements from finite field $\mathbb{F}_q$. We also assume that linear network coding is enabled in this network. To illustrate how network coding works, let us consider the case that one node s needs to deliver a series of packets $x_i, \ldots, x_h$ to a set of sinks

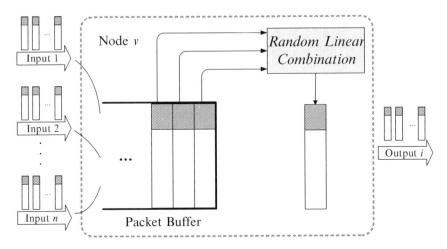

Fig. 4.2 An illustration of random linear network coding at intermediate nodes

$T \subset V$. Define the matrix of source packets as $X = [\boldsymbol{x}_1^T, \ldots, \boldsymbol{x}_h^T]^T$, i.e., X consists of all source packets as its rows. For simplicity, let $\Gamma^-(s)$ consist of h imaginary links, $\tilde{e}_1, \ldots, \tilde{e}_h$, with $y(\tilde{e}_i) = \boldsymbol{x}_i$. Then for any $e \in \Gamma^+(v), v \notin T$, $y(e)$ is calculated by linearly combining the incoming packets of v as

$$y(e) = \sum_{e' \in \Gamma^-(v)} \beta_{e'}(e)y(e') = \boldsymbol{\beta}(e)[y^T(e')]_{e' \in \Gamma^-(v)}^T \qquad (4.1)$$

where the coefficients $\beta_{e'}$ are chosen over $\mathbb{F}_q$, and the row vector $\boldsymbol{\beta}(e) = [\beta_{e'}]_{e' \in \Gamma^-(v)}$ is termed as the *local encoding vector (LEV)* of link e. By induction, $y(e)$ can be represented as the linear combination of source packets:

$$y(e) = \sum_{i=1}^{h} g_i(e)\boldsymbol{x}_i = \boldsymbol{g}(e)X \qquad (4.2)$$

where $\boldsymbol{g}(e) = [g_1(e), \ldots, g_h(e)]$ can be calculated recursively using Eq. (4.1) and is termed as the *global encoding vector (GEV)* of link e. Assume that h packets $y(e_1), \ldots, y(e_h)$ are received by a sink node v from links $e_1, \ldots, e_h$. Then, by applying Eq. (4.2), we have

$$Y = \begin{bmatrix} y(e_1) \\ \vdots \\ y(e_h) \end{bmatrix} = \begin{bmatrix} \boldsymbol{g}(e_1) \\ \vdots \\ \boldsymbol{g}(e_h) \end{bmatrix} X = GX \qquad (4.3)$$

where G is termed as the *global encoding matrix (GEM)* of node v. Since G is invertible with high probability when q is sufficiently large [13], v can reconstruct source messages X by calculating $X = G^{-1}Y$.

In practice, the source prefixes each packet x_i with the ith unit vector $\boldsymbol{u}_i$:

$$[\boldsymbol{u}_i, \boldsymbol{x}_i] = [\underbrace{0, \ldots, 0}_{i-1}, 1, \underbrace{0, \ldots, 0}_{h-i}, x_{i,1}, \ldots, x_{i,l}] \qquad (4.4)$$

where each $\boldsymbol{u}_i$ is termed as a *tag*. With the same coding operations performed on these tags, each packet will automatically contain its GEV.

4.2.2 Adversary Model

Informally, the adversary considered in this chapter aims at intercepting packets and decoding them to harvest meaningful information. It can act as an *external eavesdropper* to monitor network links, and/or as an *internal eavesdropper* to compromise intermediate nodes and read their memories (Fig. 4.3). For any eavesdropping attack W, let $E' \subset E$ denote the set of links being monitored, and $V' \subset V$ denote the set of nodes being compromised. We characterize the attack W as the set of packets intercepted by the adversary:

$$W = \{y(e) : e \in E' \cup \Gamma_v^- \cup \Gamma_v^+, v \in V'\} \qquad (4.5)$$

Then, an adversary can be defined as a set $A = \{W_i\}$, i.e., it can launch any attack belonging to A. Let W_i be a matrix whose rows contain all linearly independent GEVs of packets in W_i. Let k_i be the number of rows in W_i. Then the *capability* of the adversary can be defined as $k = \max_i k_i$. We say A is *k-capable* if it has capability k, and we say A is *global* if $k = h$. With the adversary model given above, we consider the following three different levels of security for network-coded systems:

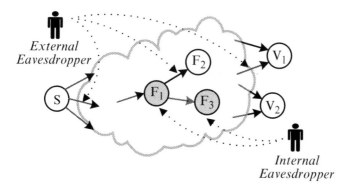

Fig. 4.3 External and internal eavesdroppers in the network

1. *Shannon security* [24]: The system is said to be Shannon secure (perfectly secure), if the adversary cannot get any information about the source messages X from the intercepted packets, which can be formulated as

$$H(X|W_i) = H(X), \forall W_i \in A \tag{4.6}$$

2. *Weak security* [2]: If no meaningful information about the source messages X can be derived from the packets intercepted by adversary, the system is said to be weakly secure, which can be formally stated as

$$H(x_i|W_i) = H(x_i), \forall x_i \in X; \forall W_i \in A \tag{4.7}$$

The difference between Shannon security and weak security can be illustrated using the following simple example: Suppose the eavesdropper *Eav* has intercepted one bit $a \oplus b$, where a and b are two i.i.d bits from the source. Then *Eav* obtains one bit of information about a and b, and the system is clearly not Shannon secure. However, *Eav* cannot recover either a or b, i.e., no meaningful information is leaked about either a or b. The system is said to be weakly secure.

3. *Computational security* [18]: Computational security is based on the assumption that the adversary is resource bounded. It is satisfied if the amount of effort to recover any meaningful information about $\forall x_i \in X$ using the best currently known methods exceeds computational resources of the adversary.

Remarks In this chapter, we will not consider Shannon security, as it is only achievable under ideal assumption that the adversary can only monitor a limited number of links [3]. In other words, Shannon security cannot be achieved when there are global eavesdroppers. As for weak security, we will show that given that the finite field size is sufficiently large and the adversary is less than h-capable, network coding is inherently weakly secure with high probability. Computational security will be our main focus, as it can be achieved using cryptographic approaches.

4.3 Security Analysis of Network Coding

In this section, we will demonstrate the weak security property of network coding through two theorems. The first one states that under certain assumptions, network coding is inherently weakly secure with high probability, while the second considers the smart adversary which can guess some combinations of the source messages.

We consider the random linear network coding (RLNC) model [13], where coding coefficients are chosen randomly from finite field $\mathbb{F}_q$.

Theorem 4.1. *For sufficiently large value of q, the probability that the adversary of capability $k < h$ will not get any meaningful information can be approximated as*

$$P_{ws}(k) = \prod_{i=1}^{k}(1 - hq^{i-h} + hq^{i-h-1}) \qquad (4.8)$$

Proof. Consider the adversary launches an attack $W_i \in A$. There are two necessary conditions for the adversary not to get any meaningful information: (1) k_i is less than the number h of source messages; otherwise, all information can be recovered, and (2) no unit vector can be derived by performing Gaussian eliminations on W_i, since a unit vector indicates that one message can be directly obtained. Since $k < h$, we have $k_i \leq k < h$, i.e., Condition (1) holds. Then we only need to evaluate the probability that Condition (2) holds when $k_i = k$.

Let $M(i,j), 0 \leq i \leq j$, denote the number of all $j \times h$ matrices Λ defined on finite filed $\mathbb{F}_q$ satisfying: a) The first i rows of Λ are linearly independent and already fixed, while the last $j - i$ rows are variable; b) No unit vector can be derived by performing Gaussian eliminations on Λ. Then, it easily follows that $M(0, k)$ is the number of all $k \times h$ matrices satisfying Condition (2).

By conditioning on whether the $(i + 1)$th row is linearly independent of the first i rows, we have the following recurrence relation about $M(i,j)$:

$$M(i,j) = q^i \cdot M(i,j-1) + (N_{i+1} - q^i) \cdot M(i+1,j) \qquad (4.9)$$

where $N_i = q^h - q^{i-1}(q-1)h$ is the number of choices for the ith row to satisfy Condition (2); q^i is the number of choices for the $(i + 1)$th row to be linearly dependent on the first i rows.

As q is sufficiently large, we have $M(i,j) \approx N_{i+1}M(i+1,j)$ which leads to the approximation $M(0,k) \approx \prod_{i=1}^{k} N_i$. Then, the probability that Condition (2) holds can be calculated as

$$P_{ws}(k) = \frac{M(0,k)}{q^{hk}} \approx \frac{\prod_{i=1}^{k} N_i}{q^{hk}} = \prod_{i=1}^{k}(1 - hq^{i-h} + hq^{i-h-1})$$

The theorem is thus proven.

We show our approximate result (i.e., Theorem 4.1) and that of Bhattad's [2] in Fig. 4.4. For comparison, we also include the accurate $P_{ws}(k)$ using the value of $M(0, k)$ calculated by Eq. (4.9). From Fig. 4.4, we can see that *if the size of finite field is sufficiently large, the probability of weak security can be made arbitrarily high*. Moreover, our approximate result is closer to the accurate one, compared to the result given by Bhattad's.

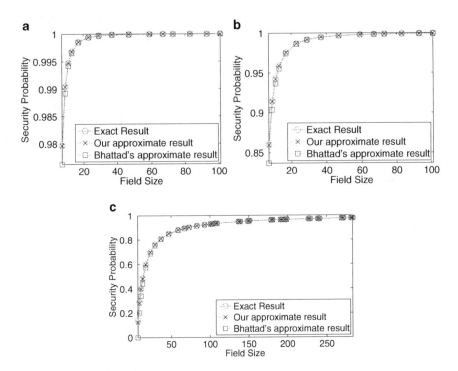

Fig. 4.4 The relationship between security probability and finite field size. (**a**) $h = 7, k = 4$, (**b**) $h = 7, k = 5$, (**c**) $h = 7, k = 6$

Next we consider a more general case where a smart adversary can accurately guess some linear combinations of source messages. The adversary is also allowed to choose the linear coefficients for the combinations. In a successful case for the adversary, it can solve more than g messages with only g guesses. We then evaluate the probability for network coding to resist this guessing threat.

Theorem 4.2. *The probability that the smart adversary can only solve g messages by g guesses is*

$$P_{gws}(k, g) = 1 - |\cup_{1 \le t \le h-g} G_t|/q^{hk}, (1 \le g < h - k) \tag{4.10}$$

where $G_t = \{W_i : \exists\{r_1, \ldots, r_t\} \subset (span(I_1, \ldots, I_{g+t}) \cap span(W_i))\}$, each $I_i, (1 \le i \le h)$ is a unit row vector of dimension h, and $I_i \ne I_j$ for $i \ne j$.

Proof. From the analysis of linear dependence, the smart adversary can recover $g + t \le h$ messages by g guesses if and only if there is a matrix W_i containing t row vectors that are linear combinations of some $g + t$ different unit row vectors.

By defining the set of all such W_i as G_t, the probability that only g messages can be recovered is calculated by eliminating the probability contributed by all G_t, $(1 \leq t \leq h - g)$.

4.4 P-Coding: The Proposed Scheme

This section defines *permutation encryption*, based on which we introduce P-Coding, a lightweight encryption scheme. Then, we introduce an enhanced scheme to further improve the security of P-Coding.

4.4.1 Permutation Encryption

We formalize the concept of permutation encryption as a special case of the classic transposition cipher [18].

Notations We term a sequence π containing each element of set $1, \ldots, n$ once and only once as a *permutation with length n*. Let $\pi(i)$ be the ith element of π, then the product of two permutations π_1 and π_2, defined by $\pi_1 \circ \pi_2$, or $\pi_1\pi_2$, is calculated using $\pi_1\pi_2(i) = \pi_1(\pi_2(i))$. Let π^{-1} be the inverse of π with respect to product operation.

Definition 4.1. Let $m = [m_1, m_2, \ldots, m_n]$ be a sequence of symbols from finite field $\mathbb{F}_q$ and k be a permutation with length n, then the *permutation encryption function (PEF)* on m using key k is defined as

$$E_k(m) = [m_{k(1)}, m_{k(2)}, \ldots, m_{k(n)}] \tag{4.11}$$

Similarly, we can define the *permutation decryption function* on c using key k as $D_k(c)$, satisfying $D_k(E_k(m)) = m$. Here, the permutation k is termed as the PEF key.

Property Since the permutation encryption only involves reordering the sequences to be encrypted, without altering any of their symbols, it is exchangeable with the linear combinations over finite field $\mathbb{F}_q$, as shown below:

- Addition: $E_k(m + n) = E_k(m) + E_k(n)$
- Scalar Multiplication: $E_k(t \cdot m) = t \cdot E_k(m)$

Remarks Note that permutation encryption is quite simple and vulnerable to cryptographic analysis [8]. However, we try to use it on top of network coding to generate considerable confusion to the adversary. It may work in the context of network coding, as packets in network coding are linear combinations of original packets. To decode it, we need GEVs. Randomly permuting the packet symbols can make the eavesdropper unable to locate the GEVs and thus fail to decode the packets. Detailed proof will be given in Sect. 4.5.

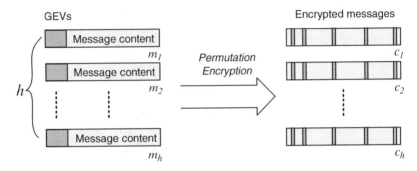

Fig. 4.5 An illustration of permutation encryption on coded messages

4.4.2 The P-Coding Scheme

The basic idea of P-Coding is to perform permutation encryptions on coded messages, as shown in Fig. 4.5. After PEF operations, symbols of the messages and corresponding GEVs can be mixed and reordered together. We will show in Sect. 4.5 that such PEF operations can generate considerable confusions to the adversary.

The P-Coding scheme primarily consists of three stages: source encoding, intermediate recoding, and sink decoding. Without loss of generality, we assume that there is a key distribution center (KDC) responsible for symmetric key establishment, so that the source and sinks can share a PEF key k at the bootstrap stage of P-Coding.

Source Encoding Consider the situation that a source s has h messages, denoted by column vectors $x_1, \ldots, x_h$, to be sent out. It first prefixes these h messages with their corresponding unit vectors, according to Eq. (4.4). Then the source performs linear combinations on these messages with randomly chosen LEVs. For instance, with LEV $\beta(e_i)$ of output link e_i, we can get the coded message $y(e_i) = [\beta(e_i), \beta(e_i)X]$, where $X = [x_1^T, \ldots, x_h^T]^T$. Finally, the source performs permutation encryption on each message $y(e_i)$ to get its ciphertext $c[y(e_i)] = E_k[y(e_i)]$.

Intermediate Recoding Since the symbols of messages and corresponding GEVs are rearranged via PEF, and the intermediate nodes have no knowledge of the key being used, it is rather difficult for them to reconstruct source messages. On the other hand, as permutation encryptions are exchangeable with linear combinations, intermediate recoding can be transparently performed on the encrypted messages:

$$c[y(e_i)] = c[\sum_{e' \in \Gamma^-(v)} \beta_{e'}(e)y(e')] = \sum_{e' \in \Gamma^-(v)} \beta_{e'}(e)c[y(e')]$$

Note that this transparency property makes P-Coding rather efficient, since no extra effort is needed at any intermediate node.

Sink Decoding For each sink node, on receiving a message $c[y(e_i)]$ from its incoming link $e_i \in \Gamma^-(v)$, it decrypts the message by performing permutation decryption on it:

$$D_k\{c[y(e_i)]\} = E_k^{-1}\{E_k[y(e_i)]\} = y(e_i) \tag{4.12}$$

Once h linearly independent messages $y(e_1), \ldots, y(e_h)$ are collected, the sink derives the following matrix representation similar to Eq. (4.3):

$$Y = \begin{bmatrix} y(e_1) \\ \vdots \\ y(e_h) \end{bmatrix} = \begin{bmatrix} \boldsymbol{g}(e_1), \boldsymbol{g}(e_1)X \\ \vdots \\ \boldsymbol{g}(e_h), \boldsymbol{g}(e_h)X \end{bmatrix} = [G, GX] \tag{4.13}$$

Finally, the source messages can be recovered by applying Gaussian eliminations on Y:

$$Y = [G, GX] \xrightarrow[\text{elimination}]{\text{Gaussian}} [I, X] \tag{4.14}$$

4.4.3 The Enhanced P-Coding Scheme

In practical network coding applications (e.g., distributed content distribution [11]), the source may need to transmit a large volume of data D. In this case, the source should first divide D into generations:

$$D = [\underbrace{\boldsymbol{x}_1, \ldots, \boldsymbol{x}_h}_{G_1}, \ldots, \underbrace{\boldsymbol{x}_{(n-1)h+1}, \ldots, \boldsymbol{x}_{nh}}_{G_n}, \ldots]$$

Then D is sent as a stream of generations, with network coding only performed among messages belonging to the same generation. In P-Coding, if the same PEF key is used throughout the transmission, single generation failure may occur, in which an accidental key disclosure in one generation will compromise the secrecy of the following transmission.

We address this problem by *randomly perturbing the key used in each generation*. More specifically, for each generation G_i, let the PEF key be used in G_i as k_i. Before each generation of data transmission, the source S conducts the following three steps: (1) S chooses a random permutation ω_i of length n, termed as *the perturbing key*; (2) S updates k_i, using the equation $k_i = \omega_i \circ k_{i-1}$, where $\circ$ denotes the product of two permutations; and (3) S encrypts ω_i using another cryptographic approach (e.g., AES [7]) and sends the ciphertext of ω_i to all sinks who can similarly update k_i.

If the perturbing key ω_i is randomly chosen each generation and communicated securely between the source and sinks, this scheme can effectively prevent the single generation failure. However, the scheme will also inevitably incur some space overhead as the perturbing key should be transmitted in each generation. One possible implementation is to prefix each packet of the ith generation with the ciphertext of ω_i. Considering that each perturbing key is of length n, the same with a tagged packet, this scheme will incur 100 % space overhead if no extra measure is taken, clearly not feasible. In the following, we will show how to make this scheme more efficient.

Definition 4.2. Suppose π is a permutation with length n, if $\pi(i) = i$ holds for each $i \notin [s, s + m - 1] \subseteq [1, n]$, we say that π is m-partial.

For a partial permutation, some elements of it are in their original positions. It can be seen that an m-partial permutation with length n can be represented by an integer $s \in [0, n - m + 1]$ and a permutation with length m. Thus, we can decrease the length of the key to m, by using an m-partial permutation as the perturbing key.

Next, we consider compressing the m-partial permutation to an integer $d \in [0, m! - 1]$ for efficient transmission. To achieve this, we must find a one-to-one correspondence between integers and permutations, so that given an integer it is efficient to calculate the corresponding permutation. Therefore, we introduce the following proposition.

Proposition 4.1. *There is a one-to-one correspondence between integers $n \in [0, m! - 1]$ and permutations π with length m.*

Proof. From basic combinatorics, any $n \in [0, m! - 1]$ can be uniquely represented as

$$n = a_{m-1}(m - 1)! + a_{m-2}(m - 2)! + \ldots, + a_1 \cdot 1! \qquad (4.15)$$

where $a_i \in [0, i]$ can be calculated using two recursive formulas: $a_i = n_i \% (i + 1)$ and $n_{i+1} = \lfloor n_i / (i + 1) \rfloor$, with initial condition $n_1 = n$. Construct a sequence $b_1, \ldots, b_{m-1}$ from $a_1, \ldots, a_{m-1}$, with $b_i = m - a_{m-i}$, and we have $b_i \in [i, m]$. Define a permutation $\omega = (1, 2, \ldots, m)$, and perform m rounds of operations: in the ith round, exchange the elements of $\omega(i)$ and $\omega(b_i)$. Then, the resultant ω is the corresponding permutation of length m. Since the above construction is a one-to-one correspondence, the proposition is proven.

Based on this proposition, we propose Algorithm 1, which aims to perturb the key using five parameters, of which k is the current PEF key; n denotes the length of the tagged packet; m denotes the partiality of the perturbing key; and s and d are chosen randomly from their respective domains to represent the perturbing key.

In the enhanced P-Coding scheme, we can employ symmetric encryptions to secure the transmission of perturbing key (s, d) from the source to sinks. Another possible approach is to let the source and sinks share a common pseudo random number generator (PRNG), so that the perturbing key (s, d) can be generated by the source and sinks in a distributed manner.

Algorithm 1: Key Perturbing Function

Input: a permutation k of length n, integers n, m, s, d with $1 \le m \le n$, $s \in [0, n - m + 1]$ and $d \in [0, m! - 1]$

Output: a perturbed permutation $\tilde{k}$ of length n

    `// to generate the sequence` $(a_1, \dots, a_{m-1})$

1 **foreach** $i \in [1, m - 1]$ **do**
2 | $a(i) \leftarrow d \% (i + 1)$;
3 | $d = \lfloor d/i + 1 \rfloor$;
4 **end**

    `// to generate the sequence` $(b_1, \dots, b_{m-1})$

5 **foreach** $i \in [1, m - 1]$ **do**
6 | $b(i) \leftarrow m - a(m - i)$;
7 **end**

    `// Initialization`

8 **foreach** $i \in [1, n]$ **do**
9 | $\omega(i) \leftarrow i$;
10 **end**

    `// to calculate the partial permutation`

11 **foreach** $i \in [1, m - 1]$ **do**
12 | $\omega(s - 1 + i) \leftrightarrow \omega(s - 1 + b(i))$;
13 **end**

    `// to perturb the current key k using` ω

14 **foreach** $i \in [1, n]$ **do**
15 | $\tilde{k}(i) \leftarrow \omega(k(i))$;
16 **end**
17 **return** $\tilde{k}$;

4.5 Security Analysis

In this section, we will analyze the security property of the proposed P-Coding scheme, both theoretically and with experimental validation. We will show the condition for permutation encryption to be secure based on elementary probability models and demonstrate that P-Coding provides a relatively high level of confidentiality. The experimental validation confirms our analysis that P-Coding is much more secure than the naive transposition cipher.

4.5.1 Probability Model

We represent the message to be encrypted as a random vector $M = [M(1), \dots, M(n)]$ over finite field $\mathbb{F}_q$. Similarly, we represent the PEF key and corresponding ciphertext as $K = [K(1), \dots, K(n)]$ and $C = [C(1), \dots, C(n)]$, respectively. For the sake of notations, we define equivalent events $\{K = k\} = \bigcap_{i=1}^{n} \{K(i) = k(i)\}$, $\{M = x\} = \bigcap_{i=1}^{n} \{M(i) = x_i\}$, and $\{C = x\} = \bigcap_{i=1}^{n} \{C(i) = x_i\}$. To simplify our analysis, assume the field size q is sufficiently large, so that the sequence M will not include duplicate symbols, i.e., $P(M(I) = M(J)) = P(I = J)$, where I and J are random variables distributed over $[1, n]$.

Definition 4.3. We say a permutation encryption is *forward random*, or has the property of *forward randomness*, if and only if it satisfies

$$P(\bigcap_{i=1}^{n}\{C(i) = x_{k(i)}\}|M = x) = 1/n!, \quad (\forall k, \forall x) \tag{4.16}$$

Similarly, we say a permutation is *backward random*, or has the property of *backward randomness*, if and only if it satisfies

$$P(\bigcap_{i=1}^{n}\{M(i) = x_{k(i)}\}|C = x) = 1/n!, \quad (\forall k, \forall x) \tag{4.17}$$

Of these two random properties of permutation encryption, backward randomness means that the plaintext *could have been* any possible order/sequence of the ciphertext with equal probability. This can make the cryptanalysis on permutation encryption degrade into exhaustive search, which promises a very strong security for permutation encryption. In the following, we will give sufficient conditions for the property of forward and backward randomness, respectively.

Theorem 4.3. *A sufficient condition for the permutation encryption to have forward randomness is $P(K = k) = 1/n!$ for each k, and K is distributed independent of M.*

Proof.

$$P(\bigcap_{i=1}^{n}\{C(i) = x_{k(i)}\}|M = x)$$

$$= \frac{P(\bigcap_{i=1}^{n}\{M(K(k^{-1}(i))) = M(i)\}, M = x)}{P(M = x)}$$

$$= \frac{P(\bigcap_{i=1}^{n}\{K(k^{-1}(i)) = i\}, M = x)}{P(M = x)} \quad (q \to \infty)$$

$$= \frac{P(\bigcap_{i=1}^{n}\{K(k^{-1}(i)) = i\})P(M = x)}{P(M = x)} \quad (\textit{Independence})$$

$$= P(K = k) = 1/n!$$

The theorem is proven according to Definition 4.3.

> **Theorem 4.4.** *A sufficient condition for the permutation encryption to have backward randomness is: the permutation encryption has forward randomness, and each $M(i) \in M$ is independently and uniformly distributed.*

Proof.

$$P(\bigcap_{i=1}^{n}\{M(i) = x_{k(i)}\}|C = x)$$

$$= \frac{P(C = x| \bigcap_{i=1}^{n}\{M(i) = x_{k(i)}\})P(\bigcap_{i=1}^{n}\{M(i) = x_{k(i)}\})}{\sum_{\pi} P(C = x| \bigcap_{i=1}^{n}\{M(i) = x_{\pi(i)}\})P(\bigcap_{i=1}^{n}\{M(i) = x_{\pi(i)}\})}$$

$$= \frac{P(\bigcap_{i=1}^{n}\{C(i) = x'_{k^{-1}(i)}\}|M = x')P(\bigcap_{i=1}^{n}\{M(i) = x_{k(i)}\})}{\sum_{\pi} P(\bigcap_{i=1}^{n}\{C(i) = x''_{\pi^{-1}(i)}\}|M = x'')P(\bigcap_{i=1}^{n}\{M(i) = x_{\pi(i)}\})}$$

$$= \frac{(1/n!)P(\bigcap_{i=1}^{n}\{M(i) = x_{k(i)}\})}{\sum_{\pi}(1/n!)P(\bigcap_{i=1}^{n}\{M(i) = x_{\pi(i)}\})} \quad (Theorem\ 3)$$

$$= \frac{\prod_{i=1}^{n} P(M(i) = x_{k(i)})}{\sum_{\pi} \prod_{i=1}^{n} P(M(i) = x_{\pi(i)})} \quad (Independence)$$

$$= q^{n}/(n! \cdot q^{n}) = 1/n!$$

The theorem is proven according to Definition 4.3.

Is P-Coding Backward Random? We claim that P-Coding is forward random since the PEF key k is generated randomly and uniformly and chosen independently of the messages to be encrypted. Then, packet in network coding undergoes rounds of random linear combinations; thus, the dependence among its elements has been largely eliminated and the distribution tends to be uniform. This fact makes P-Coding backward random to some extent according to Theorem 4.4.

Exhaustive Search is Rather Expensive Recall that each generation contains h messages, and each message has length n. To carry out exhaustive search, the adversary needs to try $O(n!)$ rounds to guess the plaintext or PEF key. In each round, it should test its guess by performing Gaussian eliminations according to Eq. (4.14), with computational complexity to be $O(h^{3})$ in terms of multiplication operations. Therefore, the computational complexity for exhaustive search is $O(n! \cdot h^{3})$.

4.5.2   Attacks on P-Coding

We consider a typical cryptanalysis on transposition cipher and evaluate its effectiveness in breaking P-Coding. This cryptanalysis is based on the nonuniform frequencies of n-letter combinations, known as n-grams [29], in natural languages. For example, bigram *TH* has a much higher frequency than bigram *QZ* in English. Using frequency statistics of n-grams, the fitness of a guessed permutation p can be easily accessed: first decrypt a large number of ciphertexts by permuting them with the inverse of p, and then evaluate how close the n-gram statistics of the decrypted messages are to those of the underlying language. After that, by searching in the neighborhood of p's with good fitness, we are expected to find other permutations with better fitness. This searching process continues until the key k is finally found. Recent studies show that optimization heuristics, e.g., genetic algorithms [12], simulated annealing [15], and ant colony [23], can be used to automate this searching process.

Though the above cryptanalysis is quite effective in breaking transposition ciphers, we argue that it does not work well for P-Coding, for the following reasons. (1) In P-Coding, before we can access the fitness of a permutation p, we should first decrypt generations of packets with p and then decode them with the GEVs. The decoding process requires $O(nh^2l)$ multiplications, where n is the number of generations, h is the generation size, and l is the length of a message. Thus, it is much more time-consuming to access the fitness of p in P-Coding than in transposition cipher. (2) Even a small change in the permutation p, say an exchange of two positions, will result in different GEVs, which may decode messages into quite different content. This means that even p has a good fitness, we cannot expect to find permutations with better fitness by searching in p's neighborhood.

To justify the above argument, we implement the genetic algorithm proposed in [8], and evaluate its feasibility to break our P-Coding scheme. The genetic algorithm for breaking P-Coding is shown as Algorithm 2, where the CM function is given as Algorithm 3. For comparison, we also include the performance of the algorithm to defeat traditional transposition ciphers. The metrics to compare include: (1) *success ratio*, the ratio of the number of rounds in which the key is recovered, to the total number of rounds, and (2) *recovery ratio*, the ratio of the average number of recovered positions of the key, to the total length of the key.

In our simulation, we chose a readable passage of length 1000 (in words), divide it into multiple messages, and perform P-Coding and transposition encryption on them, respectively. Note that the primary difference between P-Coding scheme and transposition ciphers is the former permutes messages after they are randomly and linearly coded, while the latter permute messages directly. For the parameters of genetic algorithm, we set the group size to 12 and the maximum round of mating and mutation to 100. We experiment by varying the length of permutation key from 10 to 19. For each case, we run the genetic algorithm 500 rounds for both P-Coding and transposition cipher. We report the results in Figs. 4.6 and 4.7.

From Fig. 4.6, we can see that genetic algorithm is effective to break transposition ciphers on passage we chose, particularly when the key length is 10. For longer key length, the success attack ratio stays above 0, meaning that this attack is still feasible. However, with P-Coding, the permutation encryption is rather resistant to this attack. This is justified by observing that the probability of successful attack becomes 0 when the key length increases to 14.

Algorithm 2: The Genetic Algorithm For Breaking P-Coding

Input: A ciphertext T, the permutation key length N
Output: A break of permutation key K
// Initialize a set of random permutations
1 PK $\leftarrow$ {} ;
2 **for** $1 < i < S$ **do**
3   p $\leftarrow$ Random_Permutation_Generator();
4   PK.add(p) ;
5 **end**
// Run for R rounds
6 **foreach** $n \in [1, R]$ **do**
7   **foreach** $i \in [0, S/2]$ **do**
8   **foreach** $j \in [i + 1, S/2]$ **do**
9   r $\xleftarrow{R}$ $[1, N]$;
             // Crossover and mutate
10   Child1 $\leftarrow$ CM(PK[i],PK[j],r) ;
11   Child2 $\leftarrow$ CM(PK[j],PK[i],r) ;
             // Add two children to PK
12   PK.add(Child1), PK.add(Child2) ;
13   **end**
14   **end**
     // Selection step
15   Sort keys by fitness from high to low ;
16 **end**
17 **return** $K=PK[1]$;

Algorithm 3: The CM Function

Input: Parent permutations $P1$ and $P2$
Input: Key length N, position pos
Output: Child permutation C
// Copy the first pos elements from P1
1 **foreach** $i \in [1, pos]$ **do**
2      $C[i] \leftarrow P1[i]$;
3 **end**
// Copy the remaining elements from P2
4 $i \leftarrow pos$;
5 **foreach** $j \in [1, N]$ **do**
6   **if** $P2[j]$ *is not in* C **then**
7          $C[i] \leftarrow P2[j]$;
8          $i \leftarrow i + 1$
9   **end**
10 **end**
// The mutation step
11 a $\xleftarrow{R}$ $[1, N]$, b $\xleftarrow{R}$ $[1, N]$;
12 Swap the elements of $C[a]$ and $C[b]$;
13 **return** C ;

Fig. 4.6 The effectiveness of genetic algorithm on transposition cipher and P-Coding (success ratio)

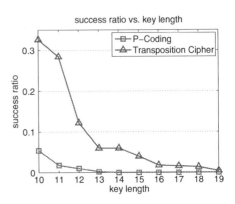

Fig. 4.7 The effectiveness of genetic algorithm on transposition cipher and P-Coding (recover ratio)

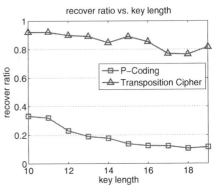

Figure 4.7 shows that even the genetic algorithm succeeds in recovering the whole key at a low probability, it can actually recover the majority of the key. This indicates the effectiveness of genetic algorithm (partially recovered key can also disclose very critical information about the plaintext). On the other hand, the recovery ratio for our P-Coding is only about 10 % (when the key length is 19), which is very low, considering that even a randomly generated sequence can have some positions that coincide with the key.

4.5.3   Security of Enhanced P-Coding

If the PEF key does not leak in any generation, the security level of enhanced scheme is as high as that of the P-Coding scheme. When single generation failure occurs, the enhanced scheme can provide two appealing properties.

Security After the compromise of security in current generation, the security level in the following ones will be strong enough to resist further attacks. We show this by evaluating the computational complexity for the adversary to guess the next PEF key based on the current one. First, it should locate the start point of key perturbing operation, which has $O(n)$ different choices. Then it should fix the correct sequence of the perturbed section of PEF key, which has $O(m!)$ different choices. It is fair to assume that these choices are equally possible, according to the randomness property of permutation encryption in P-Coding. Finally, the adversary should decode the messages by performing Eq. (4.14), which requires $O(h^3)$ multiplication operations. Thus, the computational complexity in terms of multiplication is $O(n \cdot m! \cdot h^3)$, which can be made sufficiently large by choosing m properly.

Recovery As the PEF key is perturbed randomly and incrementally, it will become more and more irrelevant to its original value with the iterations of generations. Thus, even if the current key is disclosed, its randomness to the adversary will gradually recover after several generations. Theorem 4.5 gives the numerical result to justify this argument.

Theorem 4.5. *After i generations, the expected number of all perturbed positions in the PEF key is approximately*

$$EX_i = \frac{i-1}{i+1}(n-m)[1 - (1 - \frac{m}{n-m})^{i+1}] + m \qquad (4.18)$$

when $n \to \infty$.

Proof. Suppose there is a segment L with length n. In each round, a randomly chosen subsegment of length $m < n$ in L is colored. Here n, m, and the start point of each colored subsegment are all integers. Let X_i be the total length of colored subsegments in L after i rounds, then we are expected to evaluate EX_i.

As n is sufficiently large, each start point S_i of the ith colored segments follows an independent continuous uniform distribution over $(0, \lambda)$, where $\lambda = n - m$. Define a series of random variables δ_j as

$$\delta_j = \begin{cases} S_{(1)}, & \text{if } j = 0 \\ S_{(j+1)} - S_{(j)}, & \text{if } j \geq 1 \end{cases} \qquad (4.19)$$

where $S_{(j)}$ is the jth order statistics of $S_1, \ldots, S_i$. Then it follows that $\delta_0, \delta_1, \ldots, \delta_{i-1}$ are identically distributed [22]. Define another series of random variables $\tilde{\delta}_j$, $(1 \leq j \leq i-1)$ as

$$\tilde{\delta}_j = \begin{cases} \delta_j, & \text{if } \delta_j < m \\ m, & \text{if } \delta_j \geq m \end{cases} \qquad (4.20)$$

It is evident that $\tilde{\delta}_0, \tilde{\delta}_1, \ldots, \tilde{\delta}_{i-1}$ are also identically distributed. Actually, $\tilde{\delta}_j$ is the length of segment being colored from start point $S_{(j)}$, without overlapping with that from $S_{(j+1)}$ (if there is). Thus, we have

$$EX_i = \sum_{j=1}^{i-1} E\tilde{\delta}_j + m = (i-1)E\tilde{\delta}_0 + m \qquad (4.21)$$

As the probability density function of δ_0 is

$$f(x) = (\frac{i}{\lambda})(1 - (\frac{x}{\lambda}))^{i-1}, \ (x \leq \lambda) \qquad (4.22)$$

we have

$$E\tilde{\delta}_0 = \int_0^m x dP(\tilde{\delta} < x) = \int_0^m x f(x) dx + m \int_m^\lambda f(x) dx$$

$$= \int_0^m (1 - \frac{x}{\lambda})^i dx = \frac{\lambda}{i+1}[1 - (1 - \frac{m}{\lambda})^{i+1}] \qquad (4.23)$$

Finally, the expectation of X_i can be calculated as

$$EX_i = \frac{i-1}{i+1}\lambda[1 - (1 - \frac{m}{\lambda})^{i+1}] + m \qquad (4.24)$$

The theorem is proven by replacing λ with $n - m$.

Figure 4.8 shows the approximated results from Theorem 4.5. For comparisons, we also include the exact results obtained from simulation. It can be seen that the number of perturbed positions in PEF key increases with the number of generations, meaning that its randomness will gradually recover after accidental disclosure.

4.6   Performance Evaluation

4.6.1   Analysis

The P-Coding Scheme As the PEF key can be pre-distributed at the bootstrap stage, the only online computation overhead of P-Coding comes from the permutation encryption operations at the source and decryption operations at sinks. According to Eq. (4.11), the encryption and decryption processes only involve reordering the symbols of messages, thus require $O(n)$ memory copy operations. As there are h messages in each generation, the computation overhead is then $O(n \cdot h)$ in terms of memory copy operations. Since the inherent overhead of network coding is at least $O(h^3)$ in terms of multiplication operations (due to the necessity of Gaussian eliminations), P-Coding is quite lightweight in computation. In addition, P-Coding does not cause any space overhead either.

The Enhanced P-Coding Scheme In the enhanced scheme, the source should generate two integers s and d to represent the perturbing key in each generation. It is fair to assume that the generation of these two integers can be done within a constant time. So it is the same with the encryption and decryption of them.

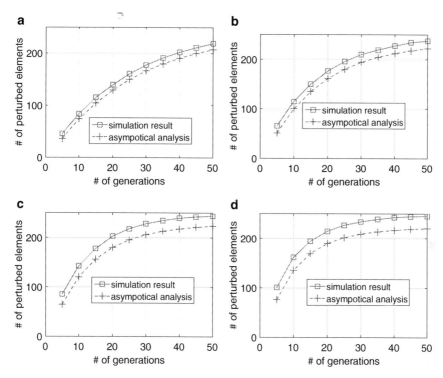

Fig. 4.8 Number of perturbed positions vs. number of generations. (**a**) $n = 255, m = 10$, (**b**) $n = 255, m = 15$, (**c**) $n = 255, m = 20$, (**d**) $n = 255, m = 25$

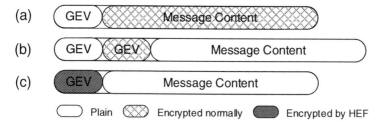

Fig. 4.9 Three cryptographic approaches for network-coded MANETs. Here GEV refers to the Global Encoding Vector

As the computational complexity of key perturbing processes is $O(n)$ according to Algorithm 1, the extra computation overhead incurred by the enhanced scheme is just $O(n)$.

Comparisons We compare the computation overhead of P-Coding and three other cryptographic schemes [9, 27] for network coding-based systems. These three schemes are depicted in Fig. 4.9: (a) the intuitive approach of directly encrypting the message content, (b) the approach proposed in [27] to encrypt only coding vectors, and (c) the approach proposed in [9] to encrypt coding vectors using HEF.

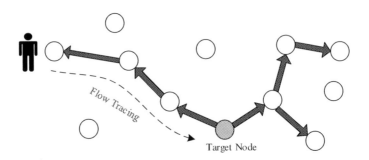

Fig. 4.10 A simple example of flow tracing

For scheme (a), source messages will be encrypted using symmetric key cryptographic algorithms (e.g., AES [7]), which cost around $O(h \cdot l)$ multiplicative operations. For scheme (b), symmetric key encryptions are only performed on GEVs, with the overhead of $O(h^2)$ multiplicative operations. In addition, this scheme also requires operations of source encoding, intermediate recoding, and sink decoding, which will cost $O(h^2)$, $O(M \cdot N \cdot h)$, and $O(h^3)$ multiplicative operations, respectively (M denotes the average number packet recoding performed by all intermediate nodes; N is the average number of combined packets for each coding operation). It also requires a space overhead of ratio $h/(n + h)$, as it inserts a duplicate GEV of length h in each packet of length n. For scheme (c), it encrypts GEVs using the public key-based Paillier cryptosystem [20], which will incur a heavy computation overhead at both source and sinks. Moreover, the linear combinations performed by intermediate nodes on GEVs will also require multiplicative and exponential operations, which are even more expensive. Since the computation overhead of P-Coding scheme is only $O(n)$ in terms of memory copy operations, and there is no space overhead, it is fair to conclude that our P-Coding scheme outperforms the other three schemes on thwarting eavesdropping attacks.

In addition, we found that P-Coding can also mitigate the problem of flow tracing, as shown in Fig. 4.10. Here, we consider three common approaches for flow tracing, i.e., size correlation, time-order correlation, and content correlation. As RLNC has messages trimmed into equal size, and buffered at intermediate nodes, it could inherently resist the first two methods. To launch message content correlation, the adversary must intercept packets of the same generation and determine if an intercepted packet in some downstream link is a linear combination of some known packets. According to [9], the computational complexity is $O(h^3 + h \cdot l)$ in terms of multiplication operations, and if some anonymous routing protocol is already in place to protect the generation number, this complexity will increase to $O(C_{wh}^h(h^3 + h \cdot l))$, where w is the total number of generations. Compared to the existing network coding with explicit GEVs, P-Coding could significantly increase the difficulty of flow tracing and thus effectively preserve the privacy of networks. This feature remarkably distinguishes P-Coding from the intuitive end-to-end encryption scheme.

Table 4.1 Comparisons among P-Coding and (a)–(c) in Sect. I

	P-Coding	(a)	(b)	(c)
Computational security	✓	✓	✓	✓
Privacy against flow tracing	✓	×	×	✓
Computation overheard	Negligible	Light	Light	Heavy
Space overhead	None	None	$h/(n+h)$	None

The above comparison is summarized in Table 4.1.

4.6.2 Experiments

In the following, we will evaluate the performance of P-Coding through experiments. For implementation of P-Coding, we first split plaintext into multiple generations of packets, then let each generation go through a random linear coding process, and perform permutation encryption on each coded packet. For comparisons, we also implement AES and 3DES (both with CBC mode) using cryptography libraries of OpenSSL [26]. Our experiment environment is a Linux desktop with 3.30GHz Intel Core i3 CPU and 4GB memory. The performance metrics we consider include encryption time, throughput, and energy consumption (Fig. 4.11).

Encryption Time We let P-Coding, AES, and 3DES encrypt a given plaintext with length up to 1K bytes and measure the time they use. The experiment setting is as follows. First, since the key size of 3DES is 192 bits (64 bits for each round), we also choose a 192-bit key for AES. Second, since 192 bits can represent a permutation of length 46, we let the P-Coding key be a random permutation of length 45. The generation size is set to be 5, meaning that each generation has 5 packets. Note that the block size of P-Coding is $(45 - 5) \times 5 = 200$ bytes.

The results are shown in Fig. 4.11a. It can be seen that the encryption time of 3DES, AES, and P-Coding increases with steps of 8 bytes, 16 bytes, and 200 bytes, respectively. In addition, the encryption time of P-Coding is around one-third that of AES. Considering OpenSSL's highly optimized implementation of AES, and our limited time in optimizing P-Coding encryption algorithms, we expect this ratio to be even lower than that.

Throughput Less encryption time means larger throughput. We will show that the throughput of P-Coding is affected by packet length (which equals the length of permutation key) and generation size (which equals the length of a GEVs). First, given a fixed generation size, the larger the packet length is, the lower the overhead of GEVs is. This is shown in Fig. 4.11b, where the generation size is set to be 5. Second, given a fixed packet length, the throughput of P-Coding decreases linearly

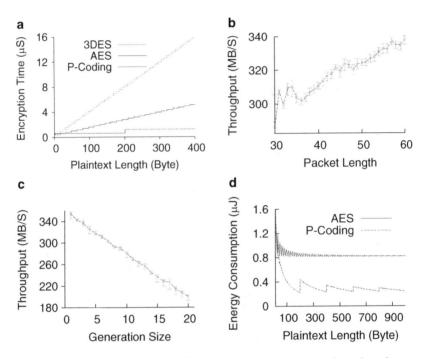

Fig. 4.11 Experimental results of P-Coding's performance (encryption time, throughput, and energy consumption). (**a**) Encryption time, (**b**) Throughput vs. packet length, (**c**) Throughput vs. generation size, (**d**) Per-byte energy consumption

with respect to the generation size, as shown in Fig. 4.11c. This is because when the packet length is fixed (here set to be 45), the increase of generation size means the increase of GEV length and thus the increase of encryption overhead.

Energy Consumption   Less encryption time also means fewer CPU cycles and less energy consumptions. To evaluate the energy efficiency of P-Coding in MANETs, we will estimate the energy consumption of P-Coding encryption on mobile nodes. Here, we choose the Motorola's "DragonBall MC68328," a common embedded microprocessor deployed in millions of wireless PDAs. Carman et al. [4] estimated that it took around $0.013mJ$ energy for AES to encrypt a 128-bit block with 128-bit key on DragonBall. Their estimation is based on the fact that it takes around 400 CPU cycles on 32-bit Intel microprocessors [1] and is obtained by scaling this result by some factor for the DragonBall microprocessors.

Here we use a similar method: we first measure the ratio of running time of P-Coding and 128-bit AES and then estimate the per-byte energy consumption of P-Coding by scaling that of AES by this ratio. Figure 4.11d gives the per-byte energy consumption of DragonBall microprocessor when using 128-bit AES and P-Coding, respectively. As the block size of P-Coding is 200 bytes, it can be seen that its per-byte energy consumption increases shapely when the plaintext length reaches multiples of 200 bytes, while drops gradually with further increase of plaintext length. As the plaintext length increases, the per-byte energy consumption of AES and P-Coding converges to 0.8 and 0.25 μJ, respectively.

4.7 Discussion

Node Mobility In the above, we have not specified how to handle node mobility. Actually, node mobility poses a challenge for key management: since network topology is constantly changing, there is no preestablished route for key establishment [5]. However, once keys are established, mobility has little impact on encryptions/decryptions. In this chapter, we view key management as an orthogonal problem, which has been studied in many previous works [19].

Extension We believe the applications of P-Coding are beyond MANETs. Any system that enables random linear network coding, like P2P live streaming [28], distributed storage [14], and file distribution [11], may use P-Coding for confidentiality. While the values of applying P-Coding in these applications are not as high as in MANETs, these applications are generally not energy constrained, and any symmetric cryptographic algorithms would function well. We will extend our scheme to other scenarios where encryption efficiency is critical.

4.8 Conclusion

This chapter studies the problem of energy saving in MANETs based on the technique of network coding. Previous studies demonstrate that network coding can reduce energy consumption with less transmissions in MANETs. In this chapter, we introduce P-Coding, a lightweight encryption scheme on top of network coding, to further reduce energy consumption in MANETs by cutting the security cost. P-Coding exploits the intrinsic security property of network coding and uses simple permutation encryptions to generate considerable confusion to eavesdropping adversaries. We show that P-Coding is efficient in computation and incurs less energy consumption for encryptions/decryptions.

References

1. Aoki, K., Lipmaa, H.: Fast implementations of aes candidates. In: Third AES Candidate Conference, pp. 13–14 (2000)
2. Bhattad, K., Narayanan, K.R.: Weakly secure network coding. In: Proceedings of International Symposium on Network Coding (NetCod) (2005)
3. Cai, N., Yeung, R.W.: Secure network coding. In: Proceedings of International Symposium on Information Theory (ISIT), p. 323 (2002)
4. Carman, D.W., Kruus, P.S., Matt, B.J.: Constraints and approaches for distributed sensor network security (final). DARPA Project report (2000)

5. Chan, A.F.: Distributed symmetric key management for mobile ad hoc networks.   In: Proceedings of IEEE INFOCOM (2004)
6. Chen, B., Jamieson, K., Balakrishnan, H., Morris, R.: Span: an energy-efficient coordination algorithm for topology maintenance in ad hoc wireless networks. Wirel. Netw. **8**(5), 481–494 (2002)
7. Daemen, J., Rijmen, V.: The design of Rijndael: AES-the advanced encryption standard. Springer, New York (2002)
8. Dimovski, A., Gligoroski, D.: Attacks on the transposition ciphers using optimization heuristics. In: Proceedings of International Scientific Conference on Information, Communication and Energy Systems and Technologies (ICEST), pp. 1–4 (2003)
9. Fan, Y., Jiang, Y., Zhu, H., Shen, X.: An efficient privacy-preserving scheme against traffic analysis attacks in network coding. In: Proceedings of IEEE INFOCOM, pp. 2213–2221 (2009)
10. Fragouli, C., Widmer, J., Boudec, J.: A network coding approach to energy efficient broadcasting: from theory to practice. In: Proceedings of IEEE INFOCOM (2006)
11. Gkantsidis, C., Rodriguez, P.R.: Network coding for large scale content distribution.   In: Proceedings of IEEE INFOCOM, pp. 2235–2245 (2005)
12. Golberg, D.E.: Genetic Algorithms in Search, Optimization, and Machine Learning. Addison Wesley, New York (1989)
13. Ho, T., Médard, M., Koetter, R., Karger, D.R., Effros, M., Shi, J., Leong, B.: A random linear network coding approach to multicast. IEEE Trans. Inf. Theory **52**(10), 4413–4430 (2006)
14. Hu, Y., Chen, H.C., Lee, P.P., Tang, Y.: Nccloud: applying network coding for the storage repair in a cloud-of-clouds. In: Proceedings of USENIX FAST (2012)
15. Kirkpatrick, S., Vecchi, M., et al.: Optimization by simulated annealing. Science **220**(4598), 671–680 (1983)
16. Li, L., Ramjee, R., Buddhikot, M., Miller, S.: Network coding-based broadcast in mobile ad-hoc networks. In: Proceedings of IEEE INFOCOM (2007)
17. Lima, L., Gheorghiu, S., Barros, J., Médard, M., Toledo, A.L.: Secure network coding for multi-resolution wireless video streaming. IEEE J. Sel. Areas Commun. **28**(3), 377–388 (2010)
18. Menezes, A.J., Van Oorschot, P.C., Vanstone, S.A.: Handbook of Applied Cryptography. CRC, Boca Raton (1996)
19. Merwe, J.V.D., Dawoud, D., McDonald, S.: A survey on peer-to-peer key management for mobile ad hoc networks. ACM Comput. Surv. **39**(1), 1 (2007)
20. Paillier, P.: Public-key cryptosystems based on composite degree residuocity classes.   In: Proceedings of EUROCRYPT (1999)
21. Potlapally, N.R., Ravi, S., Raghunathan, A., Jha, N.K.: A study of the energy consumption characteristics of cryptographic algorithms and security protocols. IEEE Trans. Mob. Comput. **5**(2), 128–143 (2006)
22. Ross, S.M.: Introduction to Probability Models. Academic, New York (2009)
23. Russell, M.D., Clark, J.A., Stepney, S.: Making the most of two heuristics: breaking transposition ciphers with ants. In: The 2003 Congress on Evolutionary Computation (2003)
24. Shannon, C.E.: Communication theory of secrecy systems.   Bell System Tech. J. **28**(4), 656–715 (1949)
25. Singh, S., Raghavendra, C., Stepanek, J.: Power-aware broadcasting in mobile ad hoc networks. In: Proceedings of IEEE PIMRC (1999)
26. The OpenSSL project. Http://www.openssl.org/ (2016)
27. Vilela, J.P., Lima, L., Barros, J.: Lightweight security for network coding. In: Proceedings of IEEE International Conference on Communications, pp. 1750–1754 (2008)
28. Wang, M., Li, B.: R2: Random push with random network coding in live peer-to-peer streaming. IEEE J. Sel. Areas Commun. **25**(9), 1655–1666 (2007)
29. Washington, L.C., Trappe, W.: Introduction to Cryptography: With Coding Theory. Prentice Hall, Englewood Cliffs, NJ (2002)
30. Wieselthier, J., Nguyen, G., Ephremides, A.: Algorithms for energy-efficient multicasting in static ad hoc wireless networks. Mob. Netw. Appl. **6**(3), 251–263 (2001)
31. Wu, Y., Chou, P.A., Kung, S.Y.: Minimum-energy multicast in mobile ad hoc networks using network coding. IEEE Trans. Commun. **53**(11), 1906–1918 (2005)

Chapter 5
Anonymous Routing for Wireless Network Coding

Wireless network coding is a promising technique that can enhance the throughput of wireless networks. However, such a technique also bears a serious security drawback: it breaks the current privacy-preserving protocols (e.g., Onion Routing), since their operations conflict each other. As user privacy in wireless networks is highly valued nowadays, a new privacy-preserving scheme that can function with wireless network coding becomes indispensable. To this end, this chapter presents a novel anonymity scheme named ANOC, which can function in network coding-based wireless mesh networks. ANOC is built upon the classic Onion Routing protocol and resolves its conflict with network coding by introducing efficient cooperation among relay nodes. Using ANOC, we can perform network coding to achieve a higher throughput, while still preserving user privacy in wireless mesh networks. We formally show how ANOC achieves the property of *relationship anonymity* and conduct extensive experiments via NSclick to demonstrate its feasibility and efficiency when integrated with network coding.

5.1 Introduction

Wireless network coding, i.e., COPE [11], is shown capable of significantly improving data throughput in wireless mesh networks [1]. In COPE, nodes operate in promiscuous mode and opportunistically perform data mixing (or coding) on the packets to be forwarded to neighboring nodes. Figure 5.1 shows three basic coding scenarios in COPE [16]. In Fig. 5.1a, node S_1 needs to send a packet P_1 to D_1, and this packet is relayed by node C; while S_2 needs to send a packet P_2 to D_2, also relayed by node C. The dashed line means that D_1 and D_2 can *overhear* P_2 and P_1, respectively, due to the broadcast nature of wireless channels. Without network coding, the communication will cost four transmissions in total: (1) S_1 sends P_1 to C, (2) C forwards P_1 to D_1, (3) S_2 sends P_2 to C, and (4) C forwards P_2 to D_2. On the

© Springer International Publishing Switzerland 2016
P. Zhang, C. Lin, *Security in Network Coding*, Wireless Networks,
DOI 10.1007/978-3-319-31083-1_5

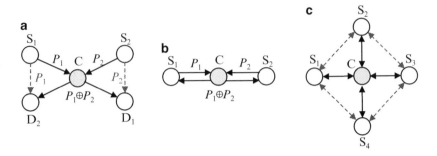

Fig. 5.1 Three typical coding scenarios of COPE, as summarized by [16]. (**a**) Coding scenario with overhearing; (**b**) Coding scenario without overhearing; (**c**) Hybrid coding scenario

other hand, with network coding, the relay node C only needs to broadcast $P_1 \oplus P_2$, and then D_1 can recover P_1 by computing $P_2 \oplus (P_1 \oplus P_2)$; D_2 can recover P_2 by computing $P_1 \oplus (P_1 \oplus P_2)$. In this way, one transmission will be saved at node C, and the network throughput can be improved. Figure 5.1b shows another possible coding scenario where no overhearing is needed; Fig. 5.1c gives a hybrid scenario that combines the former two cases.

In addition to throughput improvement, privacy preservation is also an important concern in wireless communications since (1) online privacy is highly valued by wireless users nowadays and (2) the open-air traffic in wireless medium can be easily monitored and traced. Consider, for example, a scenario where multiple clients can access a server S via a wireless mesh network. Equipped with targeted antennas, an adversary can easily intercept traffic by staying close to server S and then perform traffic analysis [2] so as to deduce the identities of users who have accessed S. Depending on the specific service provided by S, sensitive information, such as "who has accessed a web page or downloaded a file," will be disclosed. It is important to note that end-to-end encryption (e.g., SSL/TLS) only provides a limited form of privacy: while end-to-end encryption hides application payload from the adversary, the adversary can still learn the IP addresses of the client and the server in a data session.

Many techniques are proposed to provide user privacy in communication networks: Mix-Net [4, 5, 8, 9, 17, 23, 24], Onion Routing [6, 7, 12, 21], and Crowds [22] are shown to be effective in wired networks; ANODR [14], WAR [3], and Onion Ring [30] are more suitable for wireless applications. However, when a wireless network is upgraded to enable network coding, many of the above privacy-preserving protocols will not be applicable. The core reason is that the packet-mixing operations required by network coding are in *conflict* with the encryption/decryption operations required by the privacy-preserving schemes at relay nodes (details will be given in Sect. 5.3.1). Considering the rising privacy concern, as well as the increasing bandwidth demand in wireless networks, an efficient privacy-preserving scheme that can work with wireless network coding becomes highly important.

To address the above issue, this chapter introduces ANOC, i.e., anoymous network-coding-based communication for wireless mesh networks. ANOC uses Onion Routing as its building block and resolves the conflict between Onion Routing and network coding by introducing efficient cooperation (*session-key sharing* and *auxiliary decrypting*) among relay nodes. Specifically, we mainly address the following two challenges: (1) how to trigger the session key sharing in an on-demand fashion and (2) how to efficiently and securely share session keys with neighbors without leaking any information to adversaries.

With these challenges addressed, we formally show that ANOC can achieve a practical privacy requirement called *relationship anonymity* (i.e., *unlinkability* [20]), meaning that adversaries cannot associate any sender with the corresponding receiver of a data session by simply observing the wireless traffic. We also conduct extensive experiments via NSclick [18] to show that ANOC can work efficiently with network coding in wireless mesh networks.

The remainder of this chapter is organized as follows: Section 5.2 gives a formal statement of the problem to be studied. Section 5.3 motivates the basic idea of ANOC, the implementation of which is detailed in Section 5.4. Sections 5.5 and 5.6 present analytical and experimental results, respectively. Finally, Section 5.7 concludes the chapter.

5.2 Problem Statement

5.2.1 System Model

We consider a typical wireless mesh network [1] consisting of wireless routers and clients, as shown in Fig. 5.2. The routers have minimal mobility and form the infrastructure for clients. Some of these routers along the boundary of the network, termed *proxy routers*, are responsible for setting up routes for clients directly connected to them. The other routers, termed *relay routers*, reside at the core of the network and only forward packets along established paths. To enhance data throughput, wireless network coding (i.e., COPE [11]) is enabled in this mesh network: routers operate in promiscuous mode and encode/decode relayed packets opportunistically. We assume that as a basic security guarantee, end-to-end encryption (e.g., SSL/TLS) has already been deployed so that attackers cannot discover the content of packets.

5.2.2 Privacy Model

We now specify the privacy goal we aim to achieve for the system defined above. For a data session that involves communication between a *sender* (i.e., the entity that

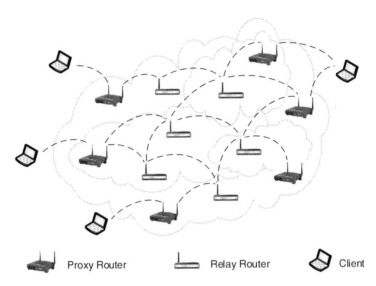

Proxy Router   Relay Router   Client

Fig. 5.2 An example of wireless mesh network

originates packets) and a *receiver* (i.e., the entity for which packets are destined), three candidate privacy models given by [20] are considered:

- *Communication Unobservability*: an adversary cannot distinguish whether a communication exists or not.
- *Sender/Receiver Anonymity*: an adversary may observe a communication session but cannot identify the sender/receiver of such a session.
- *Relationship Anonymity (or Unlinkability)*: an adversary may identify a sender or a receiver of some communication but cannot determine whether they are related or not in the same session.

Note that from the above definitions, communication unobservability offers the strongest privacy guarantee, while the unlinkability offers the weakest among the three. However, in practical systems, unobservability is mainly achieved by injecting dummy/cover packets into networks, which consumes a considerable amount of network bandwidth [27, 31]. Similar performance degradation can be observed in protocols that achieve sender/receiver anonymity. Take Crowds [22] as an example, it provides sender anonymity for web transactions. However, to hide a sender's identity, other nodes in the network need to probabilistically forward the sender's packets to each other. This will incur a large delay since each packet will traverse many more extra hops before reaching the receiver. In short, providing either communication unobservability or sender/receiver anonymity requires significant network resources and may heavily degrade the performance of legitimate applications.

On the other hand, relationship anonymity can be realized with much smaller performance degradation, which is suitable for wireless networks where bandwidth resources are generally more limited as opposed to wireline networks.

One successful scheme that achieves relationship anonymity is Onion Routing [21]. In general, relationship anonymity is sufficient for most applications that require privacy preservation, since an adversary cannot deduce the sensitive information of *"who is talking to whom,"* even though it can intercept traffic. Thus, to allow for practical deployment, in this chapter, we choose to achieve relationship anonymity for the wireless mesh network that we consider.

5.2.3 Adversary Model

Given the relationship anonymity as the privacy property to preserve, we now define the capabilities of an adversary. The adversary we consider is *passive* in nature, i.e., it *passively* monitors network traffic and will not drop, inject, or modify any packets. The only goal of the adversary is to deduce the information of "who is talking to whom" and establish the sender/receiver relationships of data sessions. To achieve this, the adversary can naively examine some identifiers (e.g., IP addresses) contained in a packet to discover the sender or receiver directly. If such identifiers are protected, the adversary can still perform traffic analysis by content correlation, size correlation, and time correlation [2].

In this chapter, we will classify the adversary into two categories:

(1) the *external adversary*, which monitors the incoming and outgoing traffic of a target node by staying close to the target node and overhearing packets via the wireless channel
(2) the *internal adversary*, which compromises and fully controls a target node and passively analyzes the traffic that traverses the target node.

We assume that the proxy routers are trustable, in the sense that they cannot be compromised by internal adversaries. This assumption is reasonable since proxy routers are mostly maintained by local network administrators to provide anonymity service to users belonging to the network, say a LAN. On the other hand, relay routers are placed in the network and managed by other parties different from the session initiators. Thus, these relay routers rather than the proxy routers have the motivation to compromise users' privacy.

5.3 ANOC: An Overview

As noted in the current literature, the main task of achieving relationship anonymity is to prevent the adversary from correlating input and output packets of relay nodes. This is commonly achieved using schemes based on Chaum's mix [4], where packets are transformed before being forwarded.

In the following, we first demonstrate the *infeasibility* of Chaum's mix-based schemes in wireless network coding and then present the design rationale of our proposed scheme. For clarity of explanation, we adopt the paradigm of *Onion Routing* [21], a classic mix-based scheme that is the core of many prior anonymous protocols [6, 7, 12, 30]. However, we emphasize that other mix-based schemes can also be used as the building block of our proposed scheme in a similar fashion.

5.3.1 Infeasibility of Onion Routing

Onion Routing [21] is an anonymous routing protocol that can achieve relationship anonymity in traditional networks without network coding. A typical Onion Routing system consists of interconnected routers called *onion routers*. Each router i is loaded with a pair of public/private keys (uk_i, rk_i) and the global knowledge of the network topology. In the following, we briefly describe how Onion Routing works when no network coding is used, using the simple cross topology shown in Fig. 5.1a.

Suppose that two end users U_1 and U_2, who are respectively connected to routers S_1 and D_1, want to set up a session. In Onion Routing, U_1 first sends a connection request to S_1. On receiving this request, S_1 determines a path to router D_1 (in this case, the path is simply $S_1 \rightarrow C \rightarrow D_1$). Then S_1 randomly selects two session keys sk_{C1} and sk_{D1} for C and D_1 respectively and constructs a layered data structure called an *onion* as $\{\{sk_{D1}, U_2\}_{uk_{D1}}, sk_{C1}, D_1\}_{uk_C}$, where uk_C and uk_{D1} are the public keys of C and D_1, respectively, and $\{\cdot\}_k$ denotes the encryption using public key k. Then S_1 sends this onion to C, which uses its private key rk_C to decrypt the onion. After decryption, C will obtain its session key sk_{C1}, the next-hop router D_1, and the embedded onion $\{sk_{D1}, U_2\}_{uk_{D1}}$. This embedded onion is then forwarded to D_1, which decrypts it using its private key to get the session key sk_{D1}. In addition, D_1 will find that it is the last hop of the route, as the next hop is the end user U_2 connected to D_1. Then D_1 forwards the connection request to U_2, and a data session is established. After the route establishment, data is transmitted using symmetric key encryptions. Specifically, using the session keys previously assigned, S_1 applies symmetric key encryption to each message M originated from U_1 and constructs $\{\{M\}_{sk_{D1}}\}_{sk_{C1}}$. Then C removes the outermost layer using sk_{C1} to get $\{M\}_{sk_{D1}}$, and finally D_1 removes the innermost layer using sk_{D1} to recover message M.

Similarly, we can apply Onion Routing for another session that uses path $S_2 \rightarrow C \rightarrow D_2$. We can assign C and D_2 session keys sk_{C2} and sk_{D2} for this session, respectively.

Suppose that network coding is enabled. We now show *how Onion Routing fails*. First, D_1 and D_2 can overhear the packets $\{\{P_1\}_{sk_{D1}}\}_{sk_{C1}}$ and $\{\{P_2\}_{sk_{D2}}\}_{sk_{C2}}$ from S_1 and S_2, respectively, and both packets will be received by C as well. Then, C will perform decryption on these two packets and get $\{P_1\}_{sk_{D1}}$ and $\{P_2\}_{sk_{D2}}$. By network coding, C would broadcast $\{P_1\}_{sk_{D1}} \oplus \{P_2\}_{sk_{D2}}$. However, *neither D_1 nor D_2 can decode the packets*, as they only overhear the packets encrypted with session keys sk_{C1} and sk_{C2} possessed by C, respectively.

Finally, it is important to note that this simple example provides an illustrative insight for larger topologies. Suppose that an adversary can eavesdrop traffic that traverses node C. When Onion Routing is used, the adversary can only tell the previous hops (i.e., S_1 and S_2) and next hops (i.e., D_1 and D_2) of node C but cannot determine the nodes that are further upstream or downstream. Such a privacy guarantee cannot be directly achieved with simple end-to-end encryption.

Summary In wireless network coding (i.e., COPE), a node can use overheard packets of other sessions to decode the packets of its own session. However, when Onion Routing is used, packets are encrypted with different session keys, and nodes cannot perform correct decoding using the stale overheard packets, which are both encoded and encrypted at the same time.

5.3.2 Design Considerations of ANOC

Cooperative networking is a relatively new design policy which encourages multiple nodes to cooperate to finish a common communication goal and is successfully applied to wireless ad hoc networks [25, 26] and content distribution networks [19]. We observe that the idea of cooperative networking can also be used here to resolve the conflicts between Onion Routing and network coding. Taking the cross topology for example again, the following two-step cooperation (as illustrated in Fig. 5.3) can help Onion Routing adapt to network coding:

1. *Session key sharing*: C shares its session key sk_{C1} and sk_{C2} with D_2 and D_1, respectively.
2. *Auxiliary decrypting*: D_2 decrypts the overheard packet $((P_1))$ using sk_{C1} to obtain (P_1) and D_1 decrypts the overheard packet $[[P_2]]$ using sk_{C2} to obtain $[P_2]$.

After the above cooperation, C can broadcast the coded packet $(P_1) \oplus [P_2]$. Then, D_1 can decode this packet to get (P_1), and decrypt (P_1) to get P_1; similarly, D_2 can obtain P_2. In this way, the conflict between network coding and Onion Routing is resolved.

Now, we consider two different approaches to achieve session key sharing between the coding node (C in the example) and the decoding nodes (D_1 and D_2

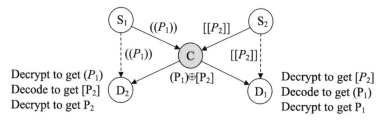

Fig. 5.3 An illustration of how relay nodes cooperate to make Onion Routing and network coding compatible. $(\cdot)$ and $[\cdot]$ denote the symmetric-key encryptions performed by S_1 and S_2 on their data packets, respectively

in the example). The first but naive approach is to simply let each router share its private key with all its one-hop neighbors (e.g., C shares rk_C with D_1 and D_2), so that when an anonymous session passing through the router is established, each of its neighbors can also obtain the session key. This approach can be carried out during the establishment phase of an anonymous session and hence will not incur any online overhead. However, the sharing of private keys would severely undermine the security of system.

For the second approach which we are going to adopt, the coding node shares its session keys (instead of private keys) with its one-hop neighbors in an on-demand fashion. Specifically, when there are coding opportunities, the router that performs coding should securely broadcast its session keys of the corresponding sessions to its neighbors. One critical point is that the share of session keys is only limited to neighboring nodes. Nodes that are further upstream or downstream cannot see the session keys; otherwise the user privacy cannot be properly preserved. Another critical point is that the key sharing procedure is only triggered when there are opportunities for network coding, so that session keys will not be shared unnecessarily.

5.4   ANOC: The Design Details

This section presents ANOC, the anonymous network coding-based communication for wireless mesh networks. ANOC is built upon the traditional Onion Routing protocol and introduces efficient cooperation (i.e., session key sharing and auxiliary decrypting) among relay nodes to resolve the conflict between Onion Routing and network coding. The technical challenges include (1) how to trigger the session key sharing in an on-demand fashion and (2) how to efficiently and securely share session keys with neighbors without leaking any information to adversaries. In the following, we show how ANOC addresses these two challenges.

Road Map of Illustration In Sect. 5.4.1, we show how to bootstrap the ANOC protocol in a given wireless mesh network; Sect. 5.4.2 describes the formats of packets to be used in ANOC; Sect. 5.4.3 shows how to set up an anonymous session with ANOC; Sect. 5.4.4 and 5.4.5 illustrate how relay nodes cooperate to enable Onion Routing to function with wireless network coding; Sect. 5.4.6 specifies how to tear down an existing anonymous session in ANOC.

5.4.1   System Setup

First, each of the routers (including proxy routers and relay routers as shown in Fig. 5.2) is assigned a unique router identifier and preloaded with a pair of public/private keys. In particular, each proxy router knows about the network topology and the public keys of all other routers in the network; each relay router only knows about its neighboring routers and their public keys. Also, each router maintains a sufficiently large buffer for bathing and reordering packets, such that the time correlation of its incoming and outgoing packets can be eliminated (see [21] for details).

In the bootstrap stage of ANOC, each router performs operations offline to establish the secure broadcast key and local neighboring table. These operations are explained below.

5.4.1.1 Secure Broadcast Key

Each router R randomly selects its *broadcast key*, which will be later used for link-layer encryptions of (1) packet headers and (2) distribution of session keys. For each of the neighboring routers, R encrypts its broadcast key using the public key of the neighbor and unicasts the ciphertext to that neighboring router. Note that this procedure has a relatively low complexity, since for mesh network consists of N nodes, each of which has an average of M neighbors, only $N \times M$ unicasts are needed. To cope with network dynamics, we also require that (1) each newly joined router exchanges its broadcast key with all of its neighbors and (2) each router in the network flushes its broadcast keys that are unused for a specific duration.

5.4.1.2 Local Neighboring Table

Each router R maintains a *local neighboring table* that records the neighbors of each of R's neighbors. The table will be used to determine whether there are coding opportunities (details will be presented in Sect. 5.4.4). For instance, for node C in Fig. 5.1c, its local neighboring table will specify that node S_1 has neighbors S_2 and S_4. The table can be easily constructed by having each router broadcast the list of its one-hop neighbors.

5.4.2 Packet Format

ANOC assumes that wireless network coding (i.e., COPE [11]) is enabled, and the packet header of COPE is placed right after the MAC header. In addition, we add a new routing header to enable anonymous routing. Figure 5.4a illustrates the layout of the COPE header and the routing header in our protocol. The routing header consists of two fields: COMMAND and CIR_ID. The COMMAND field describes the type of a packet. In ANOC, there are four types of packets:

- CONNECT (Sect. 5.4.3): for route establishment,
- DISTRIBUTE (Sect. 5.4.4): for session-key distribution,
- DATA (Sect. 5.4.5): for information delivery,
- DESTROY (Sect. 5.4.6): for tearing down an existing session.

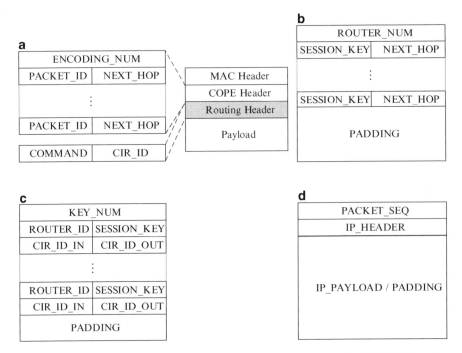

Fig. 5.4 The header and payload format of ANOC. (**a**) Header format; (**b**) CONNECT packet payload; (**c**) DISTRIBUTE packet payload; (**d**) DATA/DESTROY packet payload

The CIR_ID field carries the *circuit identifier* which enables multiple sessions to be multiplexed into a single physical channel. We explain its use in Sect. 5.4.3.

Figures 5.4b–d illustrate the four types of packets (i.e., CONNECT, DIS-TRIBUTE, DATA, and DESTROY), each of which is attached with different payload fields that are encrypted with different types of keys. We will explain each type of packets and how each is encrypted in the following subsections. In particular, we encrypt the COPE header and the routing header with the broadcast keys that have been established during the bootstrap phase (see Sect. 5.4.1). Therefore, the sensitive information such as the packet type will not be disclosed to external adversaries. Furthermore, we fix each packet to have the same size using padding so as to prevent an adversary from inferring a session through size correlation [21].

5.4.3 Session Setup

To setup a new session, the proxy router first selects the path of routers in the network toward the destination. It then generates a CONNECT packet, which specifies the selected routers and the corresponding session keys for each of the routers. Each router and its corresponding session key will be encrypted with the public key of the router, such that the CONNECT packet forms an onion structure (refer to Sect. 5.3.1).

In addition, when initiating a session, the proxy router randomly chooses a locally unique number (i.e., circuit identifier) to identify the session. This number is placed in the CIR_ID field of the CONNECT packet. When a downstream router receives the CONNECT packet, it will record the circuit identifier in the CIR_ID field and choose a new circuit identifier and replace the CIR_ID field with it. In this way, each router maintains a circuit-identifier mapping for the session. This mapping will later enable DATA packets to be routed along the path specified in the CONNECT packet.

5.4.4 Session Key Sharing

In ANOC, session keys are shared in an on-demand manner based on coding opportunities. To discover a coding opportunities, we adopt the flow-based[1] coding conditions given in [15]:

Packets of two flows F_1 and F_2 intersecting at node C can be coded together if:

(a) $D(C, F_1) \in Neig(U(C, F_2)), or D(C, F_1) = U(C, F_2)$
(b) $D(C, F_2) \in Neig(U(C, F_1)), or D(C, F_2) = U(C, F_1)$

where:

$D(C, F_i)$ denotes the *downlink node* of C in flow F_i
$U(C, F_i)$ denotes the *uplink node* of C in flow F_i
$Neig(k)$ denotes the set of all neighbors of node k

The above coding condition can be generalized to determine whether packets from $n \geq 2$ flows can be coded together, by applying this condition for each pair of flows [15]. We call flows that satisfy this condition to be *coding flows* and the node which encodes packets to be *coding node*. A router can easily judge whether packets of a newly created flow can be coded with existing flows using (1) the local neighboring table (refer to Sect. 5.4.1), (2) the identities of uplink and downlink nodes of flows passing through itself, and (3) the above coding condition. For example, in Fig. 5.1a, let $F_1 = S_1 \rightarrow C \rightarrow D_1$ and $F_2 = S_2 \rightarrow C \rightarrow D_2$. Since C knows $U(C, F_1) = S_1$, $U(C, F_2) = S_2$, $D(C, F_1) = D_1$, and $D(C, F_2) = D_2$, it can verify that $D_1 \in Neig(S_2) = \{D_1, C\}$, and $D_2 \in Neig(S_1) = \{D_2, C\}$, which implies that F_1 and F_2 are two coding flows.

After discovering the coding opportunities, the router can start sharing its session keys. Suppose that there are n coding flows $F_1, \ldots, F_n$ intersecting at router C,

---

[1]Here, a flow is equivalent to a session.

then C should distribute its session keys associated with $F_1, \ldots, F_n$ to all its one-hop neighbors. This is achieved by broadcasting a DISTRIBUTE packet encrypted with C's broadcast key, which is established in the system setup stage (refer to Sect. 5.4.1). In addition to the session keys, the DISTRIBUTE packet also contains the incoming and outgoing circuit identifiers, in order that all neighbors of the coding node can properly process overheard packets (see Sect. 5.4.5). As shown in Fig. 5.4c, a DISTRIBUTE packet contains the tuples (Router_ID, Session_Key, CIR_ID_IN, CIR_ID_OUT), where Router_ID is the identifier of the upstream router of C in this route, Session_key is the session key for C in this route, and CIR_ID_IN and CIR_ID_OUT are the incoming and outgoing circuit identifier for the session, respectively. For instance, let us consider the coding scenario in Fig. 5.3. Let the mapping of circuit identifiers in node C be $001 \rightharpoonup 255$ and $102 \rightharpoonup 123$ for sessions $S_1 \rightarrow C \rightarrow D_1$ and $S_2 \rightarrow C \rightarrow D_2$, respectively, and let C hold the session keys sk_{C1} and sk_{C2} for these two sessions, respectively. Then the two tuples contained in the DISTRIBUTE packet would be $(S_1, sk_{C1}, 001, 255)$ and $(S_2, sk_{C2}, 102, 123)$.

In our implementation, session keys are distributed incrementally. To illustrate, let us take the scenario in Fig. 5.1c. Suppose that there are initially only two anonymous sessions $S_1 \rightarrow C \rightarrow S_4$ and $S_4 \rightarrow C \rightarrow S_1$. According to our protocol, C would distribute its session keys for these two flows to its neighbors. Then S_1, S_2, S_3, S_4 will all receive and store these keys. Now suppose that a new session $S_1 \rightarrow C \rightarrow S_3$ is created. C then discovers that there are currently three coding flows. C will just distribute its session key for the new session, while the existing two sessions still use the session keys that have just been distributed.

5.4.5 Auxiliary Decrypting

In ANOC, we implement auxiliary decrypting (see Sect. 5.3.2) via a separate module named *overhearing module*, as shown in Fig. 5.5. This module consists of a key table, an overheard packet pool, and a decrypting unit. The key table stores the tuples (Router_ID, Session_Key, CIR_ID_IN, CIR_ID_OUT) of all DISTRIBUTE packets received from neighbors (see Sect. 5.4.4). When a router sends a packet or overhears a packet that is destined to a different MAC address rather than itself, it will find the router identifier of the sender (by mapping to the source MAC address) and the circuit identifier contained in the packet header. It then looks up in the key table for the tuple that has the mapping indexed by (Router_ID, CIR_ID_IN). If no tuple is found, then the overheard packet will be discarded; otherwise, if the tuple is found, then the router will decrypt the overheard packet using the session key in the tuple and replace the CIR_ID field of the overheard packet with CIR_ID_OUT in the tuple. The resulting packet will be stored in the overheard packet pool. When later the router receives a coded packet, it will look up in the overheard packet pool for packets that are necessary for decoding.

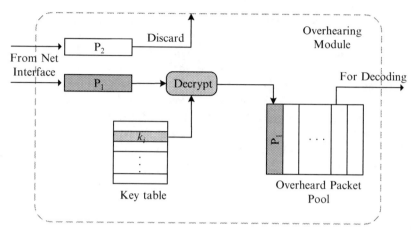

Fig. 5.5 The overhearing module in ANOC

5.4.6   Session Teardown

When an initiator needs to tear down one of its sessions, it will send a DESTROY packet along the route. Upon receiving the DESTROY packet, each router deletes all the information for the session. The neighboring routers will also expire the session keys for the session (received through DISTRIBUTE packets) after a prespecified timeout period.

5.5   Privacy Enhancement of ANOC

In this section, we analytically show how ANOC enhances privacy over a wireless mesh network that enables network coding. We consider the external adversary and the internal adversary in our adversary model defined in Sect. 5.2.3.

5.5.1   The External Adversary

We argue that ANOC can achieve relationship anonymity against the external adversary, which monitors packets via overhearing the wireless channel and attempts to perform traffic analysis via correlations of content, size, and time. We give our justifications as follows.

- *Content correlation.* In ANOC, any CONNECT, DATA, or DESTROY packet will undergo encryption or decryption when passing through each router. Thus, correlation based on content will be impossible. As for DISTRIBUTE packets, they are encrypted (using the secure broadcast key) and broadcasted without revealing the receiver identities or routing information.
- *Size correlation.* In ANOC, each packet is padded into the same size. Thus, it is impossible to perform traceback by correlating packet sizes.
- *Time correlation.* With batching and reordering operations performed at each router, a packet cannot be associated with others by examining the sending and receiving time.

5.5.2   The Internal Adversary

In Onion Routing, the whole path of a specific route is known by the involved proxy routers, while the relay routers along the route can only identify their previous and next hops. This means that if one relay router is compromised, then it cannot expose the sender/receiver relationship of the whole route. Clearly, this argument holds in ANOC as well when there are no coding opportunities (as in Onion Routing). On the other hand, when a coding node distributes its session keys using secure broadcast, its one-hop neighbors will inevitably obtain more information. This can allow the internal adversary to discover more hops in addition to its previous and next hops for a given anonymous session. In the following, we show how the internal adversary can leverage the session key sharing in our ANOC protocol to discover more hops in a session. We then give analytical results to demonstrate the number of additional hops discovered by the internal adversary is *rather limited*.

Notation Let A be the router that has been compromised and controlled by the internal adversary, and let S be a session that uses A as a relay router. Let D_i (resp. U_i) be the ith downstream (resp. upstream) node of A in session S that is ith hop away from A in the downstream (resp. upstream) direction.

Algorithm Suppose that there is another session observed at D_1 that can be coded with session S. Then, D_1's session key of S is distributed to its neighbors, according to our ANOC protocol. Algorithm 4 specifies how the adversary A can further deduce D_2.

Algorithm 4: Extra Downstream Node Discovery

---

1 Search the *Overheard Packet Pool* for a packet P that belongs to session S ;
2 Decrypt P using the session key of D_1 to get P' ;
3 Calculate the packet identifier (hash) $H(P')$ of P' ;
4 Decrypt packets sent by D_1 using D_1's broadcast key ;
5 Locate $H(P')$ in COPE headers of the packets obtained in the last step, and find the corresponding next hop of D_1 in session S (i.e., D_2) ;

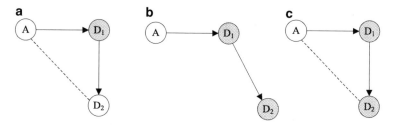

Fig. 5.6 (a),(b),(c) Three different cases of D_2. The *dashed line* indicates the neighborhood relationship, and a node is shadowed if it has a coding opportunity involving session S

Remarks The main idea of Algorithm 4 is that if D_1 is a coding node, then A can overhear packets of D_1 (which must be a one-hop neighbor of A) to determine an extra hop D_2. Now, given that A knows D_2, if D_2 is a coding node *and* a neighbor of A, then A can use the similar procedure to determine D_3, the next hop of D_2. To understand this, we consider three cases given in Fig. 5.6. In Fig. 5.6a, D_2 is a neighbor of A, but there is no coding opportunity involving S. Then, S cannot obtain the session key of S at D_2 to deduce D_3. In Fig. 5.6b, D_2 is a coding node, but D_2 is not a neighbor of A. Thus, A cannot overhear or analyze the packets sent by D_2 to deduce D_3. Only in Fig. 5.6c, D_2 is both a coding node and a neighbor of A, and A can deduce D_3 as in Algorithm 4. In general, if $D_2, D_3, \cdots, D_i$ are coding nodes and neighbors of A, then A can discover $D_3, D_4, \ldots, D_{i+1}$.

For the upstream direction, A can discover U_2 in the similar way as in Algorithm 4, i.e., by deducing whether the next hop of U_2 is U_1. The only difference is that U_2 needs to not only have a coding opportunity but also be a neighbor of A (otherwise, A cannot have the broadcast key of U_2 to decrypt its packets and analyze its next hop).

Analysis We start with the downstream direction and evaluate how many additional downstream hops (i.e., beyond D_1) that A can identify in session S established using ANOC. From the above discussion, the number of extra hops that can be identified by A depends on the placements of nodes in session S and on the number of coding opportunities of nodes in session S. Thus, we define two probabilities for a downstream node D_i in session S: (1) p_i, the probability that node D_i has a coding opportunity involving S, and (2) l_i, the probability that node D_i is a neighbor of A, conditioning on that it has a coding opportunity involving S. For simplicity, we would use the unconditional probability $q_i = p_i l_i$ in the following derivations.

The values of the two probabilities p_i and l_i depend on many factors. Intuitively, for $i \geq 2$, l_i is small (large) in a sparse (dense) network since it is less (more) likely that D_i is a neighbor of A. One observation is that the probability that a node has a coding opportunity is strictly bounded [16], and the fraction of coding traffic of COPE observed in a randomized setting is below 40 % [15]. Since $q_i \leq p_i$, we expect that both p_i and q_i are small in general.

Let $d + 1$ be the total number of downstream nodes in session S (i.e., $D_1, \cdots, D_{d+1}$). The probability that A can determine exactly n additional downstream nodes (beyond D_1) in session S (i.e., $D_2, \ldots, D_{n+1}$) is calculated by:

$$P_{n,S} = \begin{cases} p_1 \prod_{i=2}^{n} q_i (1 - q_{n+1}) & \text{if } 1 \leq n \leq d - 1 \\ p_1 \prod_{i=2}^{n} q_i & \text{if } n = d. \end{cases} \tag{5.1}$$

For $1 \leq n \leq d - 1$, $P_{n,S}$ is the product of (1) the probability that D_1 has a coding opportunity, (2) the probabilities that D_i ($2 \leq i \leq n$) have coding opportunities and are neighbors of A, and (3) the probability that D_{n+1} does not have a coding opportunity or is not a neighbor of A (so D_{n+2} cannot be deduced). For $n = d$, $P_{n,S}$ is simply the product of (1) and (2).

To simplify our analysis without losing the main message, we let $p_i = p$ and $q_i = q$ for all i for some constants p and q. Then, the expected number of additional downstream nodes (beyond D_1) that can be identified by the adversary A is

$$E_S = \sum_{n=1}^{d} n P_{n,S} = \frac{p(1 - q^d)}{1 - q}. \tag{5.2}$$

Assume $d \to \infty$, then we can bound E_S by $p/(1 - q)$. Actually, E_S would converge to this bound fast with the increase of d. This can be seen in Fig. 5.7, which plots the results of Eq. (5.2) for some values of p and l. From Fig. 5.7, we also observe that for a normal value of $l = 0.5, p = 0.4$, the expected number of additionally identified hops is less than 0.5 and even in an uncommon case where $l = 0.7, p = 0.8$, the expectation will not exceed 2. This fact justifies the limitation of adversary in discovering additional downstream hops beyond D_1 in our ANOC protocol.

Similarly we can calculate the expected number of additional upstream nodes that can be identified by the adversary A as (note that each upstream node needs to have a coding opportunity and be a neighbor of A, according to our early discussion):

$$E'_S = \frac{q(1 - q^d)}{1 - q}. \tag{5.3}$$

Note that E'_S is strictly smaller than E_S, and thus the expected number of additional upstream nodes identified by A will be less than that of downstream nodes. We are not going to plot the results of Eq. (5.3), since the curves will be much the same with that of Fig. 5.7 in shape.

5.6   Performance Evaluation

We now evaluate ANOC using realistic wireless settings. Our evaluation is based on *NSclick* [18], which embeds Click Modular Router [13] into the ns2 simulator [28]. A key design feature of nsclick is to enable a routing protocol implemented in

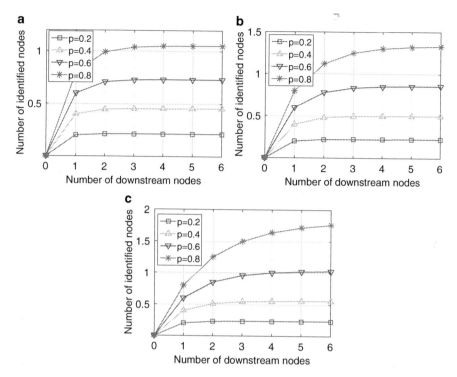

Fig. 5.7 The expected number of additional downstream nodes (beyond D_1) identified by A vs. the total number of downstream nodes (beyond D_1). (**a**) $l = 0.3$; (**b**) $l = 0.5$; (**c**) $l = 0.7$

Click to be readily deployed in a real networks with minimal configuration changes. Thus, we choose nsclick to ensure that our ANOC implementation mimics an actual system prototype used in practice.

We design two Click modules to reflect our system model: the *proxy router module*, which defines a proxy router for initiating new sessions and selecting routes, and the *relay router module*, which defines a relay router for forwarding data packets and performing network coding. For comparisons with ANOC, we also implement the following two routing protocols:

- *COPE*, the routing protocol with network coding but no anonymity protection,
- *Onion*, the Onion Routing protocol with anonymity protection but no network coding.

Our experiments focus on the following four metrics:

- *Throughput*, the aggregate throughput of all sessions in the network,
- *Coding rate*, the ratio of the number of encoded packets to the total number of forwarded packets,

- *Fairness*, the measure of how peer flows get equal throughput based on Jain's fairness index [10] $(\sum x_i)^2/(N \sum x_i^2)$ (where x_i denotes the throughput of the ith flow and N is the number of all flows),
- *Symmetric key encryption/decryption*, the computational cost of packet processing when anonymity protection is used.

The first three metrics mainly focus on the communication performance, while the last metric evaluates the computation overhead incurred by anonymous routing. In our experiments, we assume that the processing delay of packets due to symmetric key encryption/decryption is negligible. This is justified by the observation that the communication over the wireless channel is generally the performance bottleneck as opposed to the processing of symmetric key cryptographic operations.[2]

We carry out experiments using three representative topologies given in Fig. 5.8: (a) the cross topology, which is relatively simple and serves as the baseline setting; (b) the grid topology, representing a relatively complex but regular setting; and (c) the random topology, a more realistic setting to test the applicability of our scheme.

Throughout these experiments, we use 802.11b and UDP traffic sources with default settings, i.e., the transmission range is set to 250 m and the carrier sensing range is set to 550 m. For each experiment, we vary the traffic load of participating flows for the collection of results. Note that each data point that we obtain is averaged over 10 simulation runs with different random seeds.

In the following, we will first consider the throughput, coding opportunity, and fairness for each of the topologies with all three routing protocols. Then, we will measure the encryption/decryption performance specifically for Onion and ANOC.

Experiment 1. Cross Topology We first revisit the simple cross topology (see Fig. 5.8a), in which we create four sessions: $0 \rightarrow 1$, $1 \rightarrow 0$, $2 \rightarrow 3$, and $3 \rightarrow 2$, all of which are relayed by router 4. We assume that these four flows have the same offered load. Figure 5.9 shows the performance of different routing protocols. From Fig. 5.9a, we observe that when the offered load increases, the aggregate throughput achieved by ANOC also increases and is fairly close to that of COPE, while the throughput of Onion drops. The reason is that when the offered load increases, there is a higher coding opportunity for ANOC and COPE, as confirmed by Fig. 5.9b, while ANOC suffers many packet collisions under the high offered traffic load. Also, Fig. 5.9c shows that both ANOC and COPE achieve a high fairness index.

Experiment 2. Grid Topology We proceed to study the complex but regular grid topology given by Fig. 5.8b, in which there are four sessions: $5 \rightarrow 9$, $15 \rightarrow 19$, $1 \rightarrow 21$, and $3 \rightarrow 23$. We deploy the 25 nodes in a way that the radio transmission range of each node covers all neighboring nodes along its surrounding square (i.e.,

---

[2]We further conduct a benchmark test with OpenSSL [29] for 192-bit AES (a symmetric-key cryptographic algorithm) on a 2.13 GHz Intel Core 2 CPU and observe that the processing throughput is around 80 MB/s, which is much higher than the link capacity of an 802.11 wireless channel.

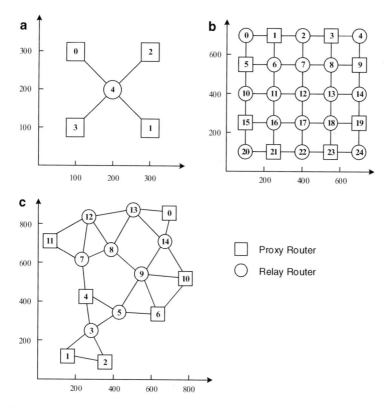

Fig. 5.8 Three topologies used in our experiments. (**a**) Cross topology; (**b**) Grid topology; (**c**) Random topology

a node can have at most eight neighboring nodes). This topology differs from the previous cross topology in that each routing path consists of more than two hops. Figure 5.10 shows the obtained results. Similar to Experiment 1, ANOC and COPE have similar performance and they both outperform Onion when the offered load is high.

Experiment 3. Random Topology We then consider a 15-node random topology where nodes are randomly placed over a plane, as shown in Fig. 5.8c. We pick five sessions for our evaluation: $0 \rightarrow 10$, $10 \rightarrow 0$, $1 \rightarrow 6$, $4 \rightarrow 2$, and $6 \rightarrow 11$. Figure 5.11 shows the results we get, which are mostly consistent with the results in previous experiments. The only difference we want to highlight is that the fairness indices for ANOC and COPE decrease when the offered load increases. The primary reason is due to the asymmetric property of this random topology, such that the sessions $0 \rightarrow 10$, $10 \rightarrow 0$, $1 \rightarrow 6$, and $4 \rightarrow 2$ receive a higher coding opportunity when the offered load increases, while the session $6 \rightarrow 11$ does not. However, this decrease in fairness is independent of the use of anonymous routing.

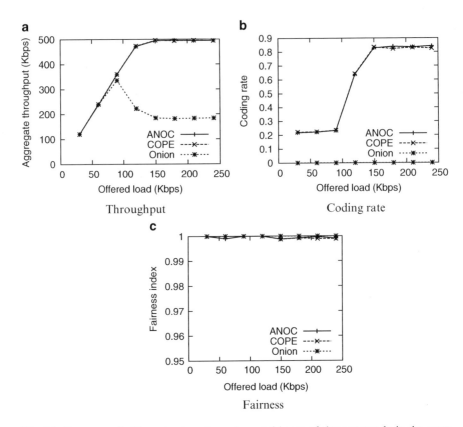

Fig. 5.9 Experiment 1: Throughput, coding rate, and fairness of three protocols in the cross topology

Experiment 4. Encryption/Decryption We finally study the performance of encryption/decryption of Onion and ANOC when anonymous routing is used. While we assume the processing delay of encryption/decryption is negligible, it remains interesting to see whether ANOC introduces significantly more encryption/decryption operations on top of network coding. Here, we examine the ratio of the total numbers of symmetric-key encryption and decryption operations to the actual number of packets successfully delivered to the destination, and the results are shown in Figs. 5.12 and 5.13.

We first look at the encryption part. Figure 5.12 shows that Onion incurs more encryptions per delivered packet than ANOC when the offered load increases. The reason is that Onion incurs more packet collisions and hence more lost packets. Encryptions performed on those lost packets will become useless. On the other hand, ANOC takes advantage of the higher coding opportunity with the increased offered load and hence reduces the number of lost packets.

On the other hand, Fig. 5.13 shows that ANOC incurs more decryptions per delivered packet than Onion, due to the auxiliary decrypting process (see Sect. 5.3).

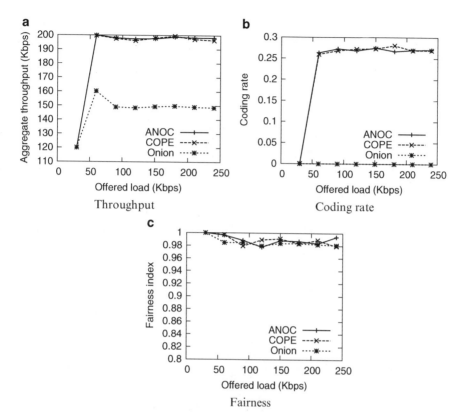

Fig. 5.10 Experiment 2: Throughput, coding rate, and fairness of three protocols for the grid topology

For instance, in the grid topology, when a node transmits a packet, up to eight of its one-hop neighboring nodes will overhear the packet and decrypt it. On the other hand, nodes in Onion will not decrypt overheard packets. Thus, we can observe the trade-off when we deploy anonymous routing on top of network coding, as there will be more decryption operations with the use of network coding.

As a brief summary, this section shows that ANOC, which combines anonymous routing and network coding, outperforms the traditional Onion Routing protocol (in terms of throughput, coding rate, and fairness), and has similar performance with existing wireless network coding scheme. Thus, ANOC maintains the improvement of the wireless network capacity as in network coding, and in the meantime, provides additional anonymity protection for wireless users.

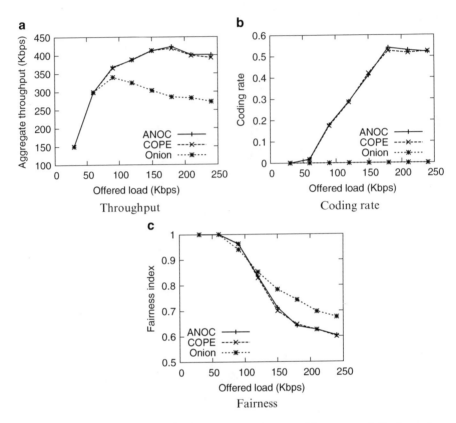

Fig. 5.11 Experiment 3: Throughput, coding rate, and fairness of three protocols for the random topology

5.7 Conclusion

This chapter introduces ANOC, a novel anonymous communication protocol which can function seamlessly with wireless network coding. The key technical feature of ANOC is that it uses Onion Routing as the building block and resolves the conflict between Onion Routing and network coding via efficient cooperation among relay nodes. We present formal analysis to show that the share of session keys in the cooperation procedure will expose a rather limited number of hops to the adversary, and thus *relationship anonymity* achievable in Onion Routing can be maintained. Furthermore, we perform extensive experiments to demonstrate that ANOC maintains the throughput, coding rate, and fairness as seen in the standard network coding paradigm. This implies that our ANOC not only keeps the advantage of network coding in the effective use of wireless network capacity, but also provides privacy for wireless users at the same time.

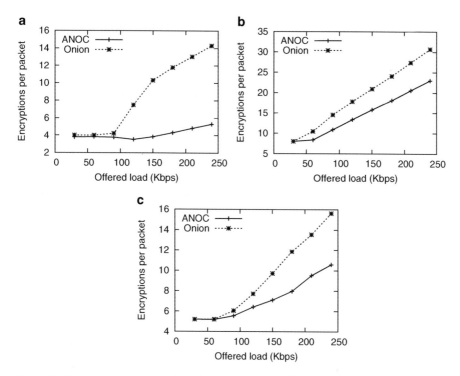

Fig. 5.12 Experiment 4: The number of encryptions per delivered packet (**a**) Cross; (**b**) Grid; (**c**) Random

References

1. Akyildiz, I.F., Wang, X., Wang, W.: Wireless mesh networks: a survey. Comput. Netw. **47**(4), 445–487 (2005)
2. Back, A., Möller, U., Stiglic, A.: Traffic analysis attacks and trade-offs in anonymity providing systems. In: Information Hiding, pp. 245–257 (2001)
3. Blaze, M., Ioannidis, J., Keromytis, A.D., Malkin, T.G., Rubin, A.: Anonymity in wireless broadcast networks. Int. J. Netw. Secur. **8**(1), 37–51 (2009)
4. Chaum, D.L.: Untraceable electronic mail, return addresses, and digital pseudonyms. Commun. ACM **24**(2), 84–90 (1981)
5. Danezis, G., Dingledine, R., Mathewson, N.: Mixminion: design of a type iii anonymous remailer protocol. In: Proceedings of IEEE Symposium on Security and Privacy (S&P), pp. 2–15 (2003)
6. DiBenedetto, S., Gasti, P., Tsudik, G., Uzun, E.: Andana: anonymous named data networking application. In: Proceedings of the Network and Distributed System Security Symposium (NDSS) (2004)
7. Dingledine, R., Mathewson, N., Syverson, P.: Tor: the second-generation onion router. In: Proceedings of the 13th USENIX Security Symposium (2004)
8. Freedman, M.J., Morris, R.: Tarzan: A peer-to-peer anonymizing network layer. In: Proceedings of ACM Conference on Computer and Communications Security (CCS), pp. 193–206 (2002)

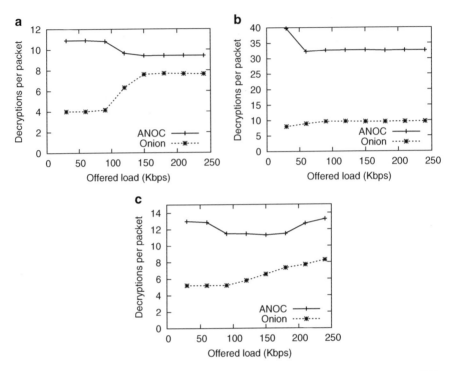

Fig. 5.13 Experiment 4: The number of decryptions per delivered packet. (**a**) Cross; (**b**) Grid; (**c**) Random

9. Gulcu, C., Tsudik, G.: Mixing e-mail with babel. In: Proceedings of the Symposium on Network and Distributed System Security (NDSS), pp. 2–16 (1996)
10. Jain, R.: The Art of Computer Systems Performance Analysis, vol. 182. Wiley, New York (1991)
11. Katti, S., Rahul, H., Hu, W., Katabi, D., Médard, M., Crowcroft, J.: Xors in the air: practical wireless network coding. ACM SIGCOMM Comput. Commun. Rev. **36**(4), 243–254 (2006)
12. Katti, S., Cohen, J., Katabi, D.: Information slicing: anonymity using unreliable overlays. In: Proceedings of the 4th USENIX Conference on Networked Systems Design & Implementation (NSDI) (2007)
13. Kohler, E., Morris, R., Chen, B., Jannotti, J., Kaashoek, M.F.: The click modular router. ACM Trans. Comput. Systems **18**(3), 263–297 (2000)
14. Kong, J., Hong, X.: Anodr: anonymous on demand routing with untraceable routes for mobile ad-hoc networks. In: Proceedings of the 4th ACM International Symposium on Mobile Ad Hoc Networking & Computing, pp. 291–302 (2003)
15. Le, J., Lui, J., Chiu, D.M.: Dcar: distributed coding-aware routing in wireless networks. In: Proceedings of the 28th International Conference on Distributed Computing Systems, pp. 462–469 (2008)
16. Le, J., Lui, J.C., Chiu, D.M.: How many packets can we encode? An analysis of practical wireless network coding. In: Proceedings of IEEE INFOCOM, pp. 371–375 (2008)
17. Möller, U., Cottrell, L., Palfrader, P., Sassaman, L.: Mixmaster protocol version 2. IETF Internet Draft, IETF (2003) https://tools.ietf.org/html/draft-sassaman-mixmaster-03

18. Neufeld, M., Schelle, G., Grunwald, D.: Nsclick user manual. University of Colorado, Boulder, CO (2003)
19. Padmanabhan, V.N., Wang, H.J., Chou, P.A., Sripanidkulchai, K.: Distributing streaming media content using cooperative networking. In: Proceedings of the 12th International Workshop on Network and Operating Systems Support for Digital Audio and Video (2002)
20. Pfitzmann, A., Hansen, M.: Anonymity, unlinkability, undetectability, unobservability, pseudonymity, and identity management-a consolidated proposal for terminology (2008). http://dud.inf.tu-dresden.de/Anon_Terminology.shtml
21. Reed, M.G., Syverson, P.F., Goldschlag, D.M.: Anonymous connections and onion routing. IEEE J. Sel. Areas Commun. **16**(4), 482–494 (1998)
22. Reiter, M.K., Rubin, A.D.: Crowds: anonymity for web transactions. ACM Trans. Inf. Syst. Secur. **1**(1), 66–92 (1998)
23. Rennhard, M., Plattner, B.: Introducing morphmix: peer-to-peer based anonymous internet usage with collusion detection. In: Proceedings of ACM Workshop on Privacy in the Electronic Society, pp. 91–102 (2002)
24. Rennhard, M., Plattner, B.: Practical anonymity for the masses with morphmix. In: Financial Cryptography, pp. 233–250 (2004)
25. Sendonaris, A., Erkip, E., Aazhang, R.: User cooperation diversity. Part I. System description. IEEE Trans. Commun. **51**(11), 1927–1938 (2003)
26. Sendonaris, A., Erkip, E., Aazhang, R.: User cooperation diversity. Part II. Implementation aspects and performance analysis. IEEE Trans. Commun. **51**(11), 1939–1948 (2003)
27. Shao, M., Yang, Y., Zhu, S., Cao, G.: Towards statistically strong source anonymity for sensor networks. In: Proceedings of IEEE INFOCOM, pp. 51–55 (2008)
28. The network simulator - ns-2. Http://www.isi.edu/nsnam/ns/ (2016)
29. The OpenSSL project. Http://www.openssl.org/ (2016)
30. Wu, X., Li, N.: Achieving privacy in mesh networks. In: Proceedings of the Fourth ACM Workshop on Security of Ad Hoc and Sensor Networks, pp. 13–22 (2006)
31. Yang, Y., Shao, M., Zhu, S., Urgaonkar, B., Cao, G.: Towards event source unobservability with minimum network traffic in sensor networks. In: Proceedings of the First ACM Conference on Wireless Network Security, pp. 77–88 (2008)

Chapter 6
Concluding Remarks and Future Directions

6.1 Concluding Remarks

Network coding is a new transmission paradigm in communication networks. Different from traditional transmission mechanisms where in-network nodes only store and forward packets, network coding allows them to perform coding on packets. Benefits of network coding include enhanced data throughput, lower energy consumption, easier node coordination, reduced bandwidth cost, etc. Up to now, network coding has been applied to various kinds of networks, e.g., wireless networks, content distribution networks, distributed storage networks, etc.

While network coding can help improve network performance, it also introduces new security and privacy issues. For example, network coding can make data transmission more vulnerable to "pollution attacks": A single illegal packet can end up polluting a bunch of good ones through the process of intermediate coding, causing receivers unable to decode properly. Besides data integrity, data confidentiality and user privacy also become quite different in the new environment of network coding. This makes existing schemes like digital signatures, encryption algorithms, and anonymity schemes either fail to work or become inefficient.

After surveying existing solutions, this book presents a series of security and privacy schemes that can fit into the new framework of network coding, based on existing cryptographic primitives and security protocols. In specific, the book has covered the following three parts:

- First, we introduced a new authentication method for network-coded systems. The method is based on the observation that the linear subspace spanned by encoded messages is invariant during transmission. Then, each message is attached a "padding", so that it is orthogonal to a randomly sampled vector. To verify the validity of a received message, a node only needs to check whether the inner product of this message and the random vector is zero. Based on this method, we introduce two concrete schemes, i.e., homomorphic

© Springer International Publishing Switzerland 2016
P. Zhang, C. Lin, *Security in Network Coding*, Wireless Networks,
DOI 10.1007/978-3-319-31083-1_6

subspace signature (HSS) and homomorphic subspace MAC (HSM). Especially, we combine their respective advantages and introduce MacSig, a novel hybrid key-based authentication scheme. We theoretically proved that MacSig can effectively and efficiently prevent pollution attacks in network-coded systems.

- Second, we theoretically analyzed the intrinsic security provided by random linear network coding. Since the intrinsic security has limitations, we design a new encryption scheme named P-Coding for network coding. P-Coding makes use of the intrinsic security property of network coding and employs simple permutation operations to provide a lightweight confidentiality. Though modeling analysis and simulated attacks, we showed that P-Coding can efficiently secure network coding-based transmissions from global eavesdroppers. The evaluation results show P-Coding can further reduce the energy consumption of encryption in mobile ad hoc networks (MANETs).

- Third, we presented a new anonymous routing protocol named ANOC, which can seamlessly integrate anonymous routing with wireless network coding. The basic idea of ANOC is to let neighboring nodes cooperate to share session keys, so that the coding operations required by network coding and the encryption/decryption operations required by Onion Routing can be performed simultaneously. ANOC is implemented in the Click modular router, simulated with NSClick. The experimental results show that the performance of ANOC is quite close to that of wireless network coding. This indicates that we can provide privacy for wireless users with anonymous routing while still maintaining a high performance with network coding.

6.2 Future Directions

There have already been a lot of works to study the security problems in network coding, while there are still some directions that have not yet received enough attention. In the following, we outline two of them.

6.2.1 The Traceback Problem in Network Coding

Distributed denial of service (DDoS) is a common and perhaps the most serious threat on today's Internet. In DDoS, the attackers can send a huge number of fake requests toward a victim server from a large number of compromised hosts distributed all around the world. As a results, normal requests from legitimate users cannot be satisfied timely. To mitigate DDoS, IP traceback techniques are developed to trace all paths followed by the DDoS requests [1, 6, 8]. Most IP traceback techniques are based on *probabilistic packet marking* and consist of two stages: packet marking and path reconstruction. In the first stage, routers probabilistically decide whether to mark a packet with their identifiers (e.g., IP addresses) or not.

In the second stage, the victim server collects enough attack packets, extracts the packet markings, and tries to recover the attack paths from them. Even functioning well in traditional IP networks, the IP traceback techniques cannot be readily applied to network coding scenarios, where DDoS is still a critical concern. For example, when a node needs to encode multiple packets (each packet carries a mark), it should decide which mark should the encoded packet carry. Thus, a research problem is how to design a good strategy that can minimize the number of packets required to construct the attack paths.

6.2.2 Malicious Node Localization for Pollution Attacks

MacSig introduced in this book is faster than its counterparts, while it can only detect the pollution attack without localizing which nodes are malicious. Siavoshani et al. [7] design such a localization method to localize malicious nodes, by leveraging the network topology features and subspace property of network coding. When there is only one malicious node, this method can localize it with uncertainty of two nodes. Wang et al. [9] propose a protocol to enable malicious node reporting. The protocol requires each node send an evidence code together with each encoded packet. Since the evidence code has the property of non-repudiation, an innocent node that detects a pollution can report the malicious node. Based on [9], Le et al. [3, 4] design a mechanism that can simultaneously support pollution detection and adversary localization. However, [3, 4, 9] all require a centralized server which knows the global network topology. Inspired by the idea of "Watch Dog" [5], Kim et al. [2] propose a method that can leverage the broadcast nature of wireless networks to localize adversaries. This method requires each node overhead packets sent by its neighbors and based on the overheard packet determine whether the neighbors are behaving well. The criterion is whether the input and output packets are belonging to the same linear subspace. The disadvantage of this approach is that it can only work for wireless networks. As a further step, we are interested in designing a distributed method for malicious node localization for both wired and wireless networks.

References

1. Dean, D., Franklin, M., Stubblefield, A.: An algebraic approach to ip traceback. ACM Trans. Inf. Syst. Secur. 5(2), 119–137 (2002)
2. Kim, M., Medard, M., Barros, J.: Algebraic watchdog: mitigating misbehavior in wireless network coding. IEEE J. Sel. Areas Commun. 29(10), 1916–1925 (2011)
3. Le, A., Markopoulou, A.: Locating byzantine attackers in intra-session network coding using spacemac. In: Proceedings of International Symposium on Network Coding (NetCod), pp. 1–6 (2010)

4. Le, A., Markopoulou, A.: Cooperative defense against pollution attacks in network coding using spacemac. IEEE J. Sel. Areas Commun. **30**(2), 442–449 (2012)
5. Marti, S., Giuli, T.J., Lai, K., Baker, M., et al.: Mitigating routing misbehavior in mobile ad hoc networks. In: Proceedings of the 6th Annual International Conference on Mobile Computing and Networking (Mobicom), pp. 255–265 (2000)
6. Savage, S., Wetherall, D., Karlin, A., Anderson, T.: Practical network support for ip traceback. In: Proceedings of ACM SIGCOMM, pp. 295–306 (2000)
7. Siavoshani, M.J., Fragouli, C., Diggavi, S.: On locating byzantine attackers. In: Proceedings of the Fourth Workshop on Network Coding, Theory and Applications (NetCod), pp. 1–6 (2008)
8. Song, D.X., Perrig, A.: Advanced and authenticated marking schemes for ip traceback. In: Proceedings of IEEE INFOCOM, pp. 878–886 (2001)
9. Wang, Q., Vu, L., Nahrstedt, K., Khurana, H.: Mis: malicious nodes identification scheme in network-coding-based peer-to-peer streaming. In: Proceedings of IEEE INFOCOM, pp. 1–5 (2010)

Printed in the United States
By Bookmasters